IMAGES
of America

TRENTON
FIREFIGHTING

Described as the biggest fire to strike Trenton since the inferno that destroyed the Buckthorn plant of John A. Roebling's Sons Company in January 1915, this general alarm blaze on February 10, 1948, in the heart of the city's business district ravaged several buildings at 120–132 South Broad Street occupied by New Jersey Floor Covering Company, D. Wolff Furniture Company, Hamilton Jewelers, A.S. Beck Shoe Company, and other businesses. A watchman in one of the buildings was killed in the fire, and two firefighters were injured. (Meredith Havens Fire Museum.)

ON THE COVER: Trenton firefighters battled this three-alarm fire at Cook Ceramic Manufacturing Company on May 27, 1954. More details of this blaze can be found on page 98. (Meredith Havens Fire Museum.)

IMAGES
of America

TRENTON
FIREFIGHTING

Michael Ratcliffe
Foreword by Chief Dennis Keenan (ret.)

ARCADIA
PUBLISHING

Published by Arcadia Publishing
Charleston, South Carolina

Printed in the United States of America

Library of Congress Control Number: 2020941285

For all general information, please contact Arcadia Publishing:
Telephone 843-853-2070
Fax 843-853-0044
E-mail sales@arcadiapublishing.com
For customer service and orders:
Toll-Free 1-888-313-2665

Visit us on the Internet at www.arcadiapublishing.com

Dedicated in loving memory to my father
James "Mike" Ratcliffe
Metuchen Volunteer Fire Department
40 Years of Service, 1965–2005
Chief of Department, 1991–1993
Line-of-Duty Death, May 3, 2005

CONTENTS

FOREWORD

I was deeply honored when Mike asked me to write a foreword for this book. Of course, I realize that I am probably the oldest firefighter he knows! I joined the Trenton Fire Department in October 1962 and was promoted to chief in 1992.

What a different department it was in the 1960s! We got two weeks of drill school that included watching the Yankees win another World Series. (Yes, they played during the day in those times.) We got our aluminum helmets—they let you know if you had gone too far into a fire by burning the tops of your ears. We rode the back step. That was fun, especially when Ed "Soup" Campbell would hit the bump at Warren and Market Streets. He would turn his head and holler, "Hang on!" The hose would lift out of the bed and our feet would be a foot above the step. Or we would ride the side of old Ladder 1, a 1938 Peter Pirsch. If you ever had to respond from Hanover Street to the state hospital up Route 29 in an early-morning sleet storm, you never would forget it!

We did have masks on the rigs (the old Burroughs filter masks), but if you were first-due you never used them. It was your job to make it as far as you could crawling on the floor. Maybe another company would relieve you with masks, but you did not count on it.

Mike was for many years a *Trenton Times* reporter covering the police and fire beat. He developed a real affinity for the fire service and is now a career firefighter himself. He has also been a valuable board member of the Meredith Havens Fire Museum of Trenton.

Mike has a wealth of knowledge about the Trenton Fire Department and shares a great deal of it in this book, giving you a look at how our department has progressed from the start in 1747 to the modern, well-equipped professional department it is now, complete with the rescue company, the honor guard, and an excellent training academy.

I surely believe you will enjoy this history of the Trenton Fire Department.

—Chief Dennis M. Keenan (ret.)

ACKNOWLEDGMENTS

Completion of this book would not have been possible without the help of a great many people. Sadly, there is not the space to describe everyone's individual contributions, so please, forgive me.

Credit, first and foremost, must go to retired Trenton fire chief Dennis Keenan. For nearly two decades, the chief gently but continuously prodded me to write a book about the Trenton Fire Department. In 2018, he invited me to join the board of trustees for the Meredith Havens Fire Museum. After I was shown the museum's archive room—an Aladdin's Cave of old photographs, annual reports, company journals, and other primary-source treasures—I knew I could no longer resist the challenge.

Special acknowledgement is also due to Laura M. Poll, the extraordinary archivist for the Trenton Free Public Library's Trentoniana Collection. Laura went out of her way to accommodate my many requests for materials and graciously allowed access to Trentoniana's fire department holdings.

Information presented in this book was culled from many sources, including the aforementioned department minutes, journals, reports, and other documents, as well as accounts published by the various newspapers that have served Trenton over the past two and a half centuries, including the *Times*, *Trentonian*, *State Gazette*, *True American*, *Trenton Federalist*, and others. John O. Raum's 1871 *History of the City of Trenton* and Francis Bazley Lee's 1895 *History of Trenton, New Jersey* were also consulted.

The fire museum's founders, Meredith Havens and Capt. Edward Gore Jr., sadly both deceased, and the late Edwin Fisher, a legendary Trenton fire dispatcher whose career spanned nearly 40 years beginning in 1927, should also be praised for their early efforts to document department history.

Most of the images in this book were scanned directly from originals preserved in the archives of the Meredith Havens Fire Museum (MHFM) and the Trentoniana Collection (TFPL). The talented photographers who took these amazing pictures and were kind enough to donate their works to these archives are also due praise. In each case where known, the original photographer is credited.

Thanks also to Steve Amiott, Harry Blaze, Ted Buriani, Justice Colucci, Shaun Dlabik, Bill Doran, Ron Dziminski, Rudy Fuessel, Anastasia Garceau, Bryan Gibbons, Tom Glover, Jeff Gore, Stephen Heaver, Steve Hodgdon, Ken Kandrac, Jack Kontura, Chris Laird, Todd Lenarski, Christine Lillpop, Jack Oakley Jr., Jack Oakley Sr., Jaime Orochena, Jim Pidcock, Mark Robotin, Rob Santello, Kevin Shea, John Smith III, Dave Smolka, Marty Sudol, Paul Tweedly, Brad Utter, Jim Yates, and members of the Society for the Preservation and Appreciation of Antique Motor Fire Apparatus in America and Fire Truck Nuts Facebook pages.

Thank you as well to my fellow fire museum board members for their confidence in me and to Katelyn Jenkins, Caroline Anderson, and everyone at Arcadia Publishing.

And most importantly, my wife, Ann, and son, James, deserve the most recognition for their unwavering patience and understanding for all those many nights and weekends over the course of 18 months that I disappeared to the library or into my home office to research and write.

I encourage anyone wanting to learn more about the Trenton Fire Department to pay a visit to the Meredith Havens Fire Museum of Trenton (trentonfiremuseum.com, 609-989-4038). Make a day of it and afterwards stop at nearby Captain Paul's Firehouse Dogs (captainpaulsdogs.com, 609-323-7253) for some great food and conversation about local fire service history from the many firefighters you'll find there.

INTRODUCTION

Volunteer firefighters had been protecting Trenton for more than four decades by the time the city was made New Jersey's capital in 1790 (actually, it was just a township back then, not becoming a city until 1792). In fact, organized firefighting here is older than the United States, predating by a generation the Revolution and George Washington's famous Christmas 1776 crossing of the Delaware River to lead the Continental Army to victory in the Battles of Trenton and Princeton.

It all started on February 7, 1747, when George Ely, Obadiah Howell, John Hunt, William Plaskett, and Thomas Tindall gathered in a blacksmith's shop at Queen (later Greene, now Broad) and Front Streets to discuss forming a fire company for Trenton. That night, it was decided that Howell would obtain buckets, fire hooks, ladders, and other equipment, while Ely and Plaskett would draft a constitution for the proposed organization. They reassembled the following evening and chose "Union" as their name, both as a nod to the successful fire company of the same name formed in 1736 by Benjamin Franklin in nearby Philadelphia and in recognition of their stated purpose to "better preserve our own and our fellow citizens' houses, goods, and estates from fire."

How often those early Trenton firefighters went to work is unknown, as the earliest records of Union Fire Company have been lost. The oldest documents known to still exist—archived in the Trenton Library's Trentoniana Collection—include a journal of meeting minutes dating to November 14, 1785, and a copy of the company's constitution from 1792. But thanks to preserved newspapers, it is known that a major blaze struck Trenton on January 30, 1772. Starting in the home of merchant Dunlap Adams and fanned by a stiff wind that sent embers showering upon neighboring roofs, the fire rapidly spread, and for a time, it was feared the entire town might be consumed. In the end, at least six dwellings and many outbuildings and stables were destroyed.

In the aftermath of that blaze, concerned Trentonians led by Rensselaer Williams met on April 2, 1772, and organized a new fire company. Taking the name Hand-In-Hand, members immediately set out equipping themselves with two leather buckets each and other tools for use in extinguishing small fires and salvaging property from blazes that could not be controlled. The Union, meanwhile, sought to improve on the bucket brigade firefighting method by purchasing its first fire engine in 1772, a small hand-tub model built in Philadelphia that was reportedly operated by just two men. A larger engine, also built in Philadelphia, was later purchased by the Union around 1786.

Another fire company, the Restoration, is said to have been formed sometime after the Union's organization but prior to the Hand-In-Hand's. No records exist for Restoration Fire Company itself, but surviving Hand-In-Hand documents show that the Restoration surrendered its engine to the Hand-In-Hand in 1779 on condition the latter repair and maintain the apparatus until such time that the Restoration should reorganize. However, a revival of the Restoration never happened.

Trenton's next fire company, known as the Resolution, was started because of a series of arsons that culminated on January 7, 1804, in the destruction of a stable and adjacent cabinet and chairmakers' shops. In hopes of catching the firebug and preventing further incendiarism, Mayor Joshua Wright offered a reward of $100—a huge sum by 1804 standards—for information leading to an arrest. Historian John Raum cites February 4, 1804, as the date of Resolution Fire Company's organization, but it may have been earlier, as a notice published by two local newspapers encouraged citizens of Trenton, Mill Hill, and Lamberton not already affiliated with the Union or Hand-In-Hand to attend a meeting on January 11, 1804, for the purpose of organizing a new fire company.

Over time, other companies were formed. Delaware Fire Company of Bloomsbury was founded on July 1, 1818, and Eagle Fire Company of Mill Hill on June 15, 1821. (Bloomsbury, Mill Hill, and Lamberton were all originally parts of Nottingham Township but merged in 1840 to create the borough of South Trenton, which in turn was annexed to become part of Trenton in 1851.)

Trenton Hose Company was created during meetings held on February 25 and March 8, 1839, in the wake of several blazes. These fires—including one that destroyed a home, a wheelwright's business, and a carpenter's shop on January 27, 1839—exposed a desperate need for more hose and improved methods of delivering water to fire scenes.

Water supply was still a major concern a decade later, when another calamity—the burning of the American Hotel's stables and other buildings at Warren and State Streets on January 18, 1848—led to the formal reorganization of Good Will Fire Company on January 24, 1848. "Our few engines were, as usual, embarrassed for want of water," the *State Gazette* wrote of the American Hotel blaze.

The date of Good Will's original organization is unknown, but it likely started in early 1839 with help from Resolution Fire Company. Minutes of the Resolution's meeting on April 4, 1839, record that "the committee appointed to dispose of the old engine have let a company of young men have the engine gratis, upon condition of its remaining under control of Resolution Fire Company." Resolution had been looking to sell its old apparatus, since its new engine—built in New York and capable of throwing water 150 feet when worked by 10 men—was purchased in 1837 for $450. Not long after being loaned the use of the old engine, the group of "young men" took to calling themselves the Good Will and helped battle a blaze that destroyed Joseph Moore's flourmill on Assanpink (as it was then spelled) Creek on August 26, 1839. Resolution members, at their next meeting, "resolved to tender to the Good Will company the thanks of this company for the energy they displayed at the recent fire at Moore's mill." What exactly happened to that first incarnation of the Good Will is unclear, but minutes of the reorganization still exist and show it was "former members" of the company who staged that revival meeting on January 24, 1848, at city hall.

The following year saw the birth of another new company and the demise of one of the old. Harmony Fire Company was created on May 9, 1849, during a meeting at Daniel Bellerjeau's home on Warren Street, above the Delaware and Raritan Canal Feeder, because organizers—noting that all the existing companies were in the city's lower half—wanted a fire engine of their own in the north. They soon began soliciting donations to buy an engine and build a firehouse. A month previous, on April 2, 1849, Resolution Fire Company disbanded and surrendered all assets to Good Will Fire Company. Interestingly, six months earlier, Resolution members acknowledged the organizations' shared history by voting to recognize the Good Will as an auxiliary of the Resolution, thereby granting Good Will members "all the privileges" of the Resolution. This action, which the Good Will felt entitled it to claim the Resolution's founding date in 1804 as its own, later became a serious bone of contention among Trenton's intensely rivalrous volunteer fire companies.

Trenton's first dedicated truck company, Protection Hook and Ladder Company, was organized on August 12, 1850, and made its first public appearance that same day when it joined the procession of firefighters and apparatus that paraded through Trenton to herald the arrival of the Good Will's newest engine (built by veteran apparatus maker John Agnew of Philadelphia). Though less than a month old, Protection Hook and Ladder reportedly did good work salvaging property from a blaze that destroyed a textile printing business on Federal Street in South Trenton on September 8, 1850. Then, after just three years, the Protection was done. A *State Gazette* account of Trenton Common Council's October 3, 1853, meeting recorded that the Protection had disbanded (no reason being given) and that the Good Will had petitioned for use of the defunct company's equipment.

Destructive fires that occurred 24 hours apart led to the creation of the Association for the Relief of Disabled Firemen in the City of Trenton, precursor to today's New Jersey State Firemen's Relief Association. About 6:30 p.m. on November 5, 1855, spontaneous combustion of oily rags ignited a fire on the fifth floor of Gaunt and Derrickson's papermill on Peace Street at Assanpink Creek. Flames rapidly spread throughout the upper stories of the building, which stood too tall for the hose streams of the hand-pumped engines to reach. Still, working through the night, firefighters managed to save valuable machinery on the lowest two floors. Exhausted firefighters went back to work the following evening after flames erupted in the stables of Samuel Kay's United States Hotel on Warren Street, near State Street. Five horses perished in that blaze, which consumed the hotel's stables and bowling alley and other adjacent structures.

In the days that followed, local newspapers were lavish in praise. "It is but due to the firemen of Trenton to say that a more active, zealous and untiring body of men is not to be found elsewhere," the *True American* opined. The *State Gazette* similarly noted, "Too much praise cannot be awarded them for their untiring and self-sacrificing zeal on both occasions, exposing themselves to the fury of the flames." In reporting that two firefighters had been badly hurt, leaving them unable to work to support their families, the *True American* remarked: "Our injured firemen should not be thrown upon the charity of the town but should be entitled to receive from the city treasury such support and compensation as their injuries may entitle them to."

While common council did not open up city coffers to aid the injured firefighters, underwriters of the companies that had insured the mill and hotel pledged $525 in recognition that their losses would have been far greater had it not been for the firefighters' herculean efforts. At a meeting presided over by Chief Engineer Alfred S. Livingston on November 12, 1855, the fire companies decided to use the insurers' generous donation to start the relief fund to aid injured firefighters and pay the funeral expenses of those who had died.

Trenton's volunteer department next grew when America Hose Company was formed on January 19, 1859. The company's first apparatus was a used hose carriage from Philadelphia. America later became an engine company in 1870 after acquiring a secondhand Amoskeag steamer.

Although the Good Will took possession of the Protection's old equipment after the hook and ladder company disbanded in 1853, Trenton's lack of a functional truck company became a topic for newspaper editorials following a destructive fire at a hay press on Stockton Street on April 12, 1864. "What has become of our hook and ladder? We understand it has been suffered to go down and the carriage to become rickety and dilapidated," the *True American* proclaimed, while the *State Gazette* professed, "The want of an efficient hook and ladder company was severely felt at the late fire. A good company, well-furnished with hooks and ladders, might have saved much property."

Finally, after several more fires, including one that gutted a shoe store, a dry goods shop, and another business near State and Greene Streets on January 1, 1866, members of Trenton Hose Company took up the challenge and announced their intention to purchase "hooks, ladders, and the necessary apparatus." On May 21, 1866, their new hook and ladder truck arrived and was celebrated with a parade that included visiting firefighters from Philadelphia. A new Trenton Hose firehouse, designed to accommodate both the company's hose carriage and the new truck, was also built and occupied in November 1866. The new rig was put to work for the first time at a fire in a pottery packing shed on Perry Street on December 19, 1866. Trenton Hose ran the apparatus until a new dedicated truck company, Washington Hook and Ladder Company, was organized on May 22, 1873, and was formally accepted into the department by common council on September 2, 1873.

Also in 1873, residents of the third and sixth wards in the southern part of the city, concerned their neighborhoods were "inadequately supplied with means to extinguish fires," organized not one but two new companies. The first was created on September 22, 1873, by younger members of Eagle Fire Company. Taking as their inspiration William Ossenberg, a veteran member of the Eagle who would later serve five terms as chief engineer, they chose the name Ossenberg Hose Company. Little is known about the other group except its name—Hibernia Hose Company—and that its petition asking to be admitted into the department came before common council on October 8, 1873. Ossenberg Hose Company, having sent its petition to council two weeks earlier and having the support of the Eagle behind it, was approved for admission into the department when council next met on October 16, 1873. No action, however, was taken on the Hibernia's petition. One council member, in opposing the Hibernia, argued that Trenton already had "too many" fire companies. The Hibernia made another application in May 1874 to join the department, but that too was tabled without action by council, and the group appears to have disbanded after that denial.

It was not until 1888, when Trenton annexed the neighboring townships of Chambersburg and Millham, that the department grew again.

Chambersburg Borough was formed from portions of Hamilton Township on April 2, 1872, and later reincorporated as a township on March 27, 1874. During its early days, Chambersburg relied on Trenton companies for fire protection. By May 1875, however, Liberty Hose Company had been formed. An old stable behind Chambersburg's town hall on South Broad Street was renovated into a firehouse, and by the end of that August, a hose carriage and 800 feet of hose had been purchased.

Liberty Hose existed but a short time. On March 8, 1876, the company's firehouse was consumed by fire. The blaze also destroyed the hose carriage and damaged the municipal building next door. Reeling from the loss of all they had worked for, company members unanimously voted to disband. Finding itself once more in need of fire protection, Chambersburg again turned to Trenton. In August 1876, Chambersburg Council approved annual payments of $75 each to the Eagle and Good Will and $40 each to Ossenberg Hose and Washington Hook and Ladder Companies for as long as Chambersburg was without its own firefighters. Creation of a fully paid department, with a proposed staff of 12, was publicly debated by Chambersburg leadership but never came to fruition.

Trenton companies continued to provide first-due fire protection to their neighbors until Mutual Steam Fire Engine Company of Chambersburg was created on June 5, 1877. (Following the arrival of the first steamers in Trenton in 1864, each company revised its legal name in a similar manner on acquiring its own new apparatus—for example, the Eagle became Eagle Steam Fire Engine Company.) Mutual remained Chambersburg's only fire company until Lincoln Hook and Ladder was formally organized on September 30, 1886. Both companies became part of the Trenton department upon Chambersburg's annexation on May 1, 1888.

Liberty Steam Fire Engine Company of Millham Township (not to be confused with the earlier short-lived Liberty Hose Company of Chambersburg) had joined the Trenton department about a month earlier when Millham—created on February 10, 1882, from the southern tip of Lawrence Township (today's Top Road and East Trenton neighborhoods)—was annexed on March 30, 1888. The company had been formally organized on May 12, 1882, following an initial meeting held on April 28 by Millham residents concerned about fire protection in their new municipality.

Thus, in its final years, the Trenton volunteer department was made up of nine engine companies, two hose companies, and two hook and ladder companies.

Throughout its 145 years of active service, the department's membership rolls boasted many of Trenton's most prominent residents, from Revolutionary War heroes and Continental Congress members to city mayors, state governors, and federal legislators. While arguments both for and against instituting a fulltime firefighting force had been published in local newspapers as far back as 1865, it was not until the volunteer companies went on strike in 1890, a year after relations with the city's board of fire commissioners soured, that the movement to create a paid department finally gained steam.

A longtime fire buff, Meredith Havens was vice president at Trenton's well-known insurance firm Walter F. Smith and Company when he and company president Rauland P. Smith began collecting local firefighting memorabilia in the 1940s. A July 1950 *Sunday Advertiser* feature reported that they had by then amassed nearly 1,000 items, many of which they displayed in their Academy Street office. These items formed the foundation of the museum that Havens and Capt. Edward R. Gore Jr. created in 1959 and opened to the public in 1961 in a small third-floor room at Fire Headquarters. The museum steadily grew so that by 1969, it had taken over headquarters' much-larger gymnasium. "Med" served as an unpaid fire and police public relations officer and did much to promote the work of emergency personnel. His colorful columns, like "With Police and Firemen" and "Meet Your Fireman," were published weekly in the *Sunday Advertiser* through the 1960s and into the mid-1970s. On April 6, 1974, the museum was renamed in his honor. Today, the Meredith Havens Fire Museum of Trenton fills the entire ground floor of the old headquarters building at 244 Perry Street. A World War II veteran, Havens died on February 9, 1979, aged 65. (MHFM.)

One

1747–1892

At first, the volunteer companies were entirely self-sufficient and supported by donations solicited directly from residents they protected and by funds generated by events like banquets and picnics. Eventually, however, the city began to help financially by making annual appropriations to cover basic operating expenses. By 1889, the various firehouses were owned and maintained by the city, but most of the engines, hose carriages, and other apparatus were still owned by the companies. Also company-owned were the horses used to pull the apparatus, and it was common practice to generate additional revenue by renting out their horses to haul local merchant wagons.

Recognizing that this often resulted in delayed responses by companies whose horses had been hired out, fire commissioners on May 29, 1889, ordered that horses be kept in their stables at all times ready for duty. Arguing that they could not continue to operate without the extra income brought in by hiring out their horses, angry volunteers demanded the city increase its annual appropriations. The commissioners refused, ignoring multiple appeals. Relations deteriorated to the point that the volunteer companies met on June 10, 1890, and voted to strike. They vowed not to answer any alarms after noon on August 9 unless city financial support was increased.

At 1:00 p.m. on August 9, 1890, the commissioners had Chief Engineer Philip Freudenmacher turn in a general alarm from Box 3 (outside city hall) of Trenton's Gamewell fire alarm telegraph. Nine companies refused to respond; only the Union, Good Will, Liberty, and Trenton Hose turned out.

The commissioners acted quickly, issuing a resolution urging common council to establish a paid department as soon as possible. Volunteers who had "remained loyal" were hired, many at $2 per day, to remain on duty as an impromptu paid department throughout the crisis.

The commissioners also found the striking companies guilty of insubordination and threatened to expel them for dereliction of duty. Realizing such punishment would prohibit them from seeking employment as paid firefighters and prevent them from drawing future benefits from the firefighters' relief fund, the strikers begged for forgiveness. After much debate, commissioners agreed the offenders would, after a 24-hour suspension, be reinstated if they promptly responded to another test alarm. That second test, conducted on August 12, was answered by all 13 companies.

The volunteers went back to work, but their days were numbered. On April 13, 1891, by more than 1,400 votes, Trentonians passed a referendum approving a paid department. The volunteers' last day on the job came a year later, April 4, 1892.

Union Fire Company took delivery of its first steamer, a third-class model manufactured by Button Fire Engine Works of Waterford, New York, on September 28, 1864. That apparatus was traded in eight years later as part of the Union's purchase of the third-class Button steamer pictured here. This new 5,200-pound engine arrived on September 2, 1872. After 100 pounds of steam pressure was built up during testing the next day, the new apparatus pumped water through 1,000 feet of hose and a 1⅛-inch nozzle to throw a solid stream a distance of 166 feet. (TFPL.)

Over the years, the Union had several homes. Among the earliest locations were North Warren Street and, later, West State Street. In June 1850, the Union moved into a new firehouse on Hanover Street shared with Trenton Hose Company. Ironically, fire damaged that building January 9, 1858. The Union relocated again in 1862 to this new structure at 16 Perry Street, which stood until it was razed and replaced on the same site by another new firehouse that was formally opened on April 5, 1888, with a celebratory parade, dedication ceremony, and banquet. (MHFM.)

Shown here and in the preceding image is the Union's 1872 Button steamer after extensive refurbishment in 1879–1880. In a demonstration on September 24, 1880, the revamped engine pumped a stream more than 250 feet high over the spire of St. Mary's Church (before it was a cathedral). This photograph was taken on April 4, 1892, before the grand parade that saluted the end of the volunteer companies' service. The building is the Union's second Perry Street firehouse, which went on to become the headquarters of the paid department. "In Union there is strength" was the company's motto. (TFPL.)

This firehouse at 29 North Willow Street was built in 1861 for Hand-In-Hand Fire Company. The Hand-In-Hand—whose motto was "Always ready"—was previously quartered in a building on Chancery Street. This hand-powered engine was built in Philadelphia by John Agnew and delivered to Trenton on June 6, 1848. The apparatus was painted dark blue and was adorned on each side with a mural of the goddess of liberty holding an American flag. Clasped hands, symbolic of the company's name, appeared below the mural. (TFPL.)

On July 1, 1851, Hand-In-Hand received a hose carriage purchased from Neptune Fire Company of Philadelphia. For several years, Hand-In-Hand members assigned to operate the hose carriage informally called themselves Neptune Hose Company but remained under the Hand-In-Hand's governance. In late 1854, Hand-In-Hand's 1848 engine went back to Philadelphia for overhaul. The refurbished apparatus returned on February 7, 1855, and remained in service until Hand-In-Hand's first steamer, built by Clapp and Jones of Hudson, New York, arrived on July 1, 1868. As with apparatus deliveries then, the steamer was welcomed with a parade two days later. (TFPL.)

Hand-In-Hand's 1868 engine was replaced by a third-class steamer built by Silsby Manufacturing of Seneca Falls, New York, that arrived in Trenton on May 18, 1878. The company operated from the same Willow Street firehouse through the final hours of the volunteer department, when this photograph was taken on April 4, 1892. This engine was the company's third steamer, another third-class Silsby, delivered on December 28, 1885. The paid department first used Hand-In-Hand's building as a storehouse, but in July 1894, after renovations, Truck 1 relocated here from the old Washington Hook and Ladder firehouse. (TFPL.)

Resolution Fire Company.

Trenton, *April 4th* 182*8*

This certifies that *E. W. Potts* is an actual member of a regular fire company, called the "Resolution Fire Company of the city of Trenton," which said company possess a fire engine, and does not consist of less than sixteen, nor more than thirty men, actual members.

Cl'k. *[signature]*   *[signature]* Pres't.

This certificate, dated April 4, 1828, confirmed Edward W. Potts as an active member of Resolution Fire Company. One of the last major fires fought by the Resolution before disbanding in 1849 was the February 2, 1847, one that gutted the Rising Sun Tavern on Warren Street. The Resolution also helped battle the March 28, 1843, blaze that destroyed the True American Inn on Broad Street in Mill Hill (which had served as George Washington's headquarters on January 1–2, 1777, during the Second Battle of Trenton). Owner Henry Katzenbach's daughter and two men died in that fire. (TFPL.)

Delaware Fire Company's 1881 Button-built steamer and 1889 Button hose carriage are shown on April 4, 1892, outside the company's South Warren Street building near Bridge Street. The company's first steamer, delivered on December 23, 1865, was also a second-class Button. Its motto was "Always ready in the hour of peril." On August 21, 1887, an arson fire gutted Delaware's stables. Weeks earlier, on July 9, another blaze of suspicious origin destroyed Union Fire Company's stables, killing a fire horse named Old Billy. (TFPL.)

The Eagle became the first fire company in Trenton to possess a steam fire engine when it took delivery on January 6, 1864, of a third-class steamer built by G.J. and J.L. Chapman of Philadelphia. This apparatus (refurbished in 1869) was replaced by a new second-class Clapp and Jones steamer that arrived on June 14, 1882. This, in turn, was traded in when the company bought a lighter Clapp and Jones steamer five years later. Tested two days after its May 14, 1887, delivery, that new third-class steamer (shown outside the Eagle's South Broad Street firehouse in both photographs) was able to throw a stream 245 feet after pumping through 250 feet of hose and a 1⅜-inch nozzle. James McCabe is in the driver's seat above around 1890. The April 4, 1892, photograph below shows Chief Engineer Philip Freudenmacher on the far right. Freudenmacher is in many of the photographs taken that day that appear in this book. The Eagle had two mottoes: "Efficient aid" and "When summoned we obey." A fire of suspicious origin destroyed the Eagle's stables on June 23, 1874, killing four horses and damaging the firehouse itself. (Above, MHFM; below, TFPL.)

Soon after starting out in 1839 with an old hose carriage borrowed from Resolution Fire Company, Trenton Hose Company purchased its own apparatus and built a small firehouse on North Warren Street opposite the Third Presbyterian Church. In 1850, it moved into a new firehouse built in partnership with Union Fire Company on Hanover Street between Warren and Greene (Broad) Streets. The two organizations shared that space until Trenton Hose sold its interests to the Union and, in June 1858, moved into its own firehouse erected nearby on Hanover Street east of Greene Street. Shown here is the building Trenton Hose remained in until the volunteer department's disbanding in 1892. It was built in 1866 on the site of the 1858 firehouse (that structure having been sold and moved by the buyer) and was specifically designed with a second apparatus bay to accommodate the new hook and ladder truck the company had purchased earlier that year. Also shown is the new hose carriage the company received on August 6, 1868. Built by Edmund Young of Philadelphia for $1,500, it was decorated with ornate gold-accented designs and polished plate glass mirrors with a monogram of the company's initials on either side. (TFPL.)

Seen here from left to right are Trenton Hose Company's 1869 officers: Assistant Foreman Joseph W. Howell, Foreman Alfred W. Packer, and Assistant Foreman John G. Bigelow. A common practice of the time saw volunteers often make "grand excursions" to other cities for parades, firefighting demonstrations, elaborate balls, banquets and, of course, sightseeing. Over the years, Trenton fire companies played host to visiting firefighters from nearby cities such as New Brunswick, Rahway, Philadelphia, and Wilmington, and those farther afield like New Haven, Connecticut, and Poughkeepsie, New York. Just two days after their 1868 hose carriage arrived, 40-plus Trenton Hose members took their new apparatus on a train ride to Massachusetts to visit Lynn and Boston. In 1874, company members, again bringing their hose carriage and accompanied by a 16-member band, made a nine-day western trip touring Cincinnati, Cleveland, Buffalo, and Niagara. (MHFM.)

In February 1876, Trenton Hose took delivery of this new carriage, built by Edward B. Leverich of New York. Originally white with gold striping, the apparatus was repainted maroon and gold in 1880. It carried 600 feet of hose. An improvement over the traditional method of connecting the apparatus alarm gong to one of the wheels so that it struck indiscriminately with every rotation, this new carriage was equipped with a spring-loaded foot pedal that allowed the driver to sound the device only when needed to clear the way or to signal some other warning. (TFPL.)

Trenton Hose purchased its first horse in 1875. Before then, company apparatus was pulled by members, a dangerous practice at the best of times. On May 1, 1872, while pulling the 1868 carriage to a fire, John G. Bigelow slipped and was run over. Fortunately, his injuries proved minor. While using horses meant firefighters no longer had to place themselves in the path of their rigs' heavy wheels, the increased speed of horse-powered responses posed new dangers. Falls and crashes happened frequently, often with deadly results, as evidenced by many line-of-duty deaths. One nonfatal example of this occurred on Christmas night in 1888, when Trenton Hose's carriage, swerving to avoid a collision with Good Will's engine while going to a fire, ran into a streetcar. Driver Lyman Hill and fellow firefighter Frank Baker were thrown off and badly hurt. Two others, George Gore and Charles H. Allen, narrowly escaped harm. The 1876 carriage seen here (the one above undated, the one below from April 4, 1892) was assigned to Engine 4 after the paid department took over. "When duty calls, 'tis ours to obey" was Trenton Hose's motto. (Both, TFPL.)

Shown here is Good Will Fire Company's first steamer. Built by Amoskeag of Manchester, New Hampshire, it arrived in Trenton on February 20, 1864, ironically one day after flames destroyed Samuel K. Wilson's woolen mill on Factory Street. This engine had a cursed existence. Firefighter Robert S. Anthony was killed when, while pulling it in the pre-horse era, he fell under its wheels as Good Will members raced the Eagle's steamer while returning from a fire on May 20, 1864. On May 18, 1868, a civilian was fatally injured when he too was run over by the engine. (MHFM.)

This is how the Good Will firehouse on South Warren Street looked on July 4, 1876. Trenton celebrated the nation's centennial that day with a 2,200-person parade, followed by a program of speeches and evening fireworks. Before moving to this location in 1862, the Good Will—"Our name is our motto"—was quartered in a firehouse that was built in 1849 on Washington (now Lafayette) Street. On October 23, 1878, a Category 2 hurricane later dubbed the "Gale of 1878" toppled Good Will's bell tower, sending its 2,000-pound bell crashing through a neighboring property's roof. (MHFM.)

This April 4, 1892, photograph shows the new third-class Amoskeag steamer delivered to the Good Will on January 16, 1881. Capable of throwing a stream 250 feet from a 1¼-inch nozzle, it weighed 5,200 pounds, had nickel plating, and was painted carmine with gold striping. Third from left, before the steamer's rear wheel, is longtime Good Will volunteer (and Civil War veteran) James Nugent. He was hired as the first captain of the paid department's Engine 1, which took over the Good Will firehouse. Nugent suffered a fatal heart attack while on duty on May 24, 1899. (TFPL.)

Though founded as a hose company, the America became an engine company in June 1870 after purchasing a second-class 1868 Amoskeag steamer from Camden. In 1886, while being shipped to New York for refurbishment, the apparatus was badly damaged in a railroad crash. It returned following a complete rebuild and was still in service when this photograph was taken on April 4, 1892, outside the America's firehouse at 522 Perry Street, opposite Ewing Street. (The firehouse was built in 1868 from the same plans used to construct the Delaware's firehouse a year earlier.) "To conquer is our aim" was the America's motto. (TFPL.)

Harmony Fire Company members can be seen in this undated photograph. Harmony's first engine, a hand-powered model built by John Agnew in Philadelphia, was delivered on January 22, 1852. It remained in service until the company received its first steamer, a second-class Amoskeag, on September 12, 1864. Construction of Harmony's first firehouse had started in May 1851 near the junction of Warren and Greene (Broad) Streets, not far from where the Trenton Battle Monument now stands. On August 26, 1851, the almost-finished building was destroyed by a raging, wind-driven fire that spread from a neighboring rope factory. Harmony rebounded by building a new firehouse on Princeton Avenue north of Pennington Avenue. A new city-built firehouse was raised for the company in 1861 on North Warren Street. Harmony added a stable in late 1865 and soon became one of the first companies in Trenton to use horses not for just parades but for regular fire duty. Firefighters briefly went back to pulling the apparatus themselves after equine influenza swept the nation in 1872 and sickened all city fire horses. On November 14, 1872, the *State Gazette* implored citizens to help pull the heavy fire engines during the horse flu epidemic. (MHFM.)

"We strive to save" was Harmony Fire Company's motto. Shown here in another of the series of photographs taken of the various companies before the volunteers' April 4, 1892, farewell parade is Harmony's second steamer, another second-class Amoskeag delivered to Trenton on October 18, 1880. Just months earlier, on June 5, 1880, the company's 1864 steamer was badly damaged when it overturned at Perry and Greene Streets while going to a fire. There were no injuries that day, but four firefighters were hurt when Harmony's hose carriage flipped over while responding to another fire on February 18, 1882. (TFPL.)

Having started using old equipment purchased from Trenton Hose Company, Washington Hook and Ladder Company on November 18, 1878, received a Leverich-built truck carrying a 64-foot chain extension ladder. This was replaced by a new $3,000 Hayes truck equipped with a 75-foot extension ladder that arrived on September 27, 1886. Washington's South Warren Street firehouse (seen here with the Hayes apparatus on April 4, 1892) opened in February 1876. The company's former home on Fair Street, having been converted into a carpenter's shop, was gutted by flames on September 8, 1876. Washington's motto was "Our name is our country's glory." (TFPL.)

Ossenberg Hose Company's first apparatus was an old carriage donated by Eagle Fire Company. In recognition of that support, and the fact that the hose company's founders had been Eagle members, Ossenberg adopted the Eagle's motto—"When summoned we obey"—as its own. Ossenberg soon purchased a new Button-built carriage. The company ran this as its front-line apparatus for many years until it went into reserve status following the June 30, 1882, delivery of another new hose carriage from Brooklyn, New York. Ossenberg's firehouse at Second and Furman Streets can be seen here. The image above is from April 4, 1892; the one below is likely from later that same day, after the volunteer department's farewell parade. Pets and horses of Trenton's fire companies—like Ossenberg's dog Dink—occasionally featured in local newspaper writeups. Dink, the *Sunday Advertiser* reported, knew many tricks but often got drunk. Also fond of beer were the Union's engine horses Major and Harry who, it was said, would drink bucketsful as their reward after making a long run to a fire. Maltese cat Joke was noted for playfully napping on the back of horse Tom in the Delaware's stables. (Above, TFPL; below, MHFM.)

Mutual Fire Company's first engine was a third-class steamer built in Newark, New Jersey, in 1878 by John N. Dennison. This was replaced by the third-class Button-built steamer (delivered on October 20, 1884) that is visible in this April 4, 1892, photograph. Sitting in the engine driver's seat is Frank Schollenberger, who joined the paid department and suffered a fatal heart attack while on duty in 1908. Also seen here is Mutual's firehouse at the corner of South Clinton Avenue and Dye Street, which it moved into in November 1878. Mutual's motto was "We help one another." (TFPL.)

On December 24, 1886, three months after organizing, Lincoln Hook and Ladder Company took delivery of a secondhand ladder truck. But it soon discovered that the rig was unfit for service, with worn, worm-eaten wheels and cracks filled with putty. It eventually acquired Washington's old Leverich truck, shown here driven by John Owens on April 4, 1892, outside Lincoln's firehouse, a very narrow building on South Clinton Avenue. Owens went on to join the paid department and died fighting a fire in 1918. "We honor the name we bear" was Lincoln's motto. (TFPL.)

Confident it would soon own its own steamer, Millham Township's newly formed firefighting organization in May 1882 chose Liberty Steam Fire Engine Company as its name. Luckily, Eagle Fire Company was anticipating the arrival of its new engine and agreed to sell its 1864 steamer to Liberty for $550. This apparatus (which had been refurbished in 1869) came into Liberty's possession in July 1882 and remained in use until Liberty's new third-class LaFrance-built steamer arrived in Trenton on July 18, 1889. The excitement of that apparatus's delivery was overshadowed days later when, while going to a fire on Taylor Street on July 24, Liberty's hose carriage accidentally knocked down a young girl, badly injuring her. The child died of her injuries on August 15, and the grief-stricken company paid her funeral expenses. Liberty, like the Mutual and Lincoln, became part of Trenton's volunteer department in 1888 after the city annexed Millham and Chambersburg. The company, whose motto was "Liberty to all," opened its new North Clinton Avenue firehouse with a housewarming banquet on December 4, 1889. The photograph above was taken on April 4, 1892. The one below is unconfirmed but was likely taken the same day. (Above, TFPL; below, MHFM.)

This was all that remained of the Third Presbyterian Church on North Warren Street following a fire on July 4, 1879, started by an Independence Day fireworks rocket landing on the roof. Flames rapidly consumed the roof and steeple (which had previously survived another, smaller fire ignited by lightning on July 2, 1874). Several firefighters, including former chief engineer Charles Yard, were working a hoseline in the church when the roof collapsed, raining down heavy timbers and shingles. They miraculously escaped with only minor injuries and burns because the high-backed pews under which they had ducked had prevented the flaming debris from crushing them to death. The era of the volunteer department is filled with many other spectacular fires—such as the blaze that burned the wooden Calhoun Street Bridge from one bank of the Delaware River to the other on June 25, 1884. So too are there many stories of the peril these early firefighters faced, like eight men who were badly injured and burned—but survived—when, while salvaging property from the first floor, the top floors of a burning storehouse collapsed on them at Hamilton Rubber Company on Meade Street, off North Clinton Avenue, on July 9, 1884. (TFPL.)

At 10:20 a.m. on January 17, 1884, an alarm of fire was sounded from Box 23, at State and Willow Streets, after smoke was seen issuing from a cupola on the New Jersey State House roof. The blaze in the attic, caused by a defective chimney flue, was quickly extinguished with minimal damage. It was a different story a year later, when this much larger fire gutted the State Street wing of the capitol. Box 23 was again sounded, this time at 2:45 a.m. on March 21, 1885. Two loud explosions occurred and flames were racing through the front of the building when firefighters arrived. Frigid morning temperatures hampered firefighting efforts, as it took 15 minutes to thaw out frozen hydrants. One firefighter lost his footing on the ice-covered roof and would have plummeted to his death had he not halted his slide by grabbing hold of a lightning rod. Before flames reached the museum on the third floor and the roof collapsed, about 20 firefighters and old soldiers, led by Adj. Gen. William S. Stryker, braved the choking smoke and rescued the battle flags that had been carried by New Jersey regiments during the Civil War. (Both, MHFM.)

State of New Jersey.
Executive Department.
Trenton, March 26th 1885

Dear Sir:

I desire in the name of the State to thank you officially in a more personal way than I was able to do in my report to the Legislature for your successful efforts in saving the colors of the New Jersey Volunteers from destruction by fire on the morning of March 21. 1885.

It will ever be a source of patriotic pride to you and your family that you were able to perform this priceless service to the State.

Truly yours –

Leon Abbett
Governor of New Jersey

Mr. Alexander Grugan,
   Trenton. –

The men who saved the Civil War colors during the state house fire were personally thanked by Gov. Leon Abbett. In this letter to one of the rescuers, Washington Hook and Ladder's Alexander Grugan, Abbett wrote: "It will ever be a source of patriotic pride to you and your family that you were able to perform this priceless service to the state." Grugan would go on to have an illustrious career with the paid department. He survived the wall collapse that killed Hoseman Charles Wood in August 1896 and, despite having had part of his left foot amputated, returned to work nine months later. On November 14, 1910, Grugan (promoted to lieutenant of Truck 1 in 1906) rescued a woman from a burning West State Street home. Scaling a ladder to the third floor, Grugan smashed a window, climbed in, and searched through blinding smoke to find and carry out the unconscious woman. Promoted to captain in 1911, Grugan was assigned as company commander of Chemical 1, then the headquarters auxiliary engine, and then finally Truck 3. He suffered a stroke, the effect of past line-of-duty injuries, while working on December 12, 1916, and died the next day. (MHFM.)

City of Trenton.
CHIEF ENGINEER'S OFFICE.

Trenton, *Dec 9th* 188 *4*

This Certifies, *that the* *Trenton*
Hose *Company put the first water on the Fire*
at *the Carriage Manufactoring Establishm*
owned by *Fitzgibbon & Crisp*
on *Tuesday* *the* *9 day of Dec* *188 4*

Edwin S. Mitchell
Chief Engineer.

Rivalry among the volunteer companies manifested in different ways—like the physical fight Eagle and Good Will members had while returning from a barn fire in Lawrence Township on August 5, 1866, for which both companies were suspended for a month by Chief Engineer John A. Weart—but a notable example of their competitiveness was how companies raced to put water on a blaze. Getting "first water" was such a big deal that the chief issued a First Water Certificate after every working fire. Shown here are certificates issued to Trenton Hose Company after destructive fires at Fitzgibbon and Crisp's carriage factory on December 9, 1884, and the slaughterhouse of John Taylor (state senator, pork roll inventor, and longtime Harmony Fire Company president) on March 26, 1887. Not having their own engine, first-arriving Trenton Hose members would often put a "plug stream" in service from a hoseline fed directly by hydrant pressure. Another source of bragging rights was being selected to respond to out-of-city fires, like when the Good Will and Delaware engines were sent, via rail, to New Brunswick on February 7, 1885, to help battle a raging fire started by the crash of an oil train. (Both, TFPL.)

City of Trenton.
CHIEF ENGINEER'S OFFICE.

Trenton, N. J. March 26 188

This Certifies, *That the* *Trenton Hose*
Company put the first water on the fire at *John Taylor*
Slauter House on Peruine Aveue
on *the 26 day of March* *1887*

Thomas Saxton *Chief Engineer.*

TRUE AMERICAN PRINT.

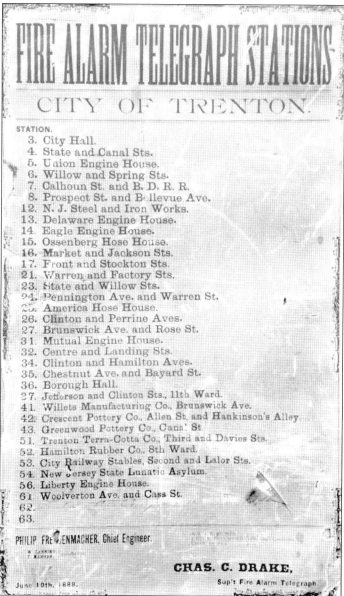

FIRE ALARM TELEGRAPH STATIONS
CITY OF TRENTON.

STATION.

3. City Hall.
4. State and Canal Sts.
5. Union Engine House.
6. Willow and Spring Sts.
7. Calhoun St. and B. D. R. R.
8. Prospect St. and Bellevue Ave.
12. N. J. Steel and Iron Works.
13. Delaware Engine House.
14. Eagle Engine House.
15. Ossenberg Hose House.
16. Market and Jackson Sts.
17. Front and Stockton Sts.
21. Warren and Factory Sts.
23. State and Willow Sts.
24. Pennington Ave. and Warren St.
25. America Hose House.
26. Clinton and Perrine Aves.
27. Brunswick Ave. and Rose St.
31. Mutual Engine House.
32. Centre and Landing Sts.
34. Clinton and Hamilton Aves.
35. Chestnut Ave. and Bayard St.
36. Borough Hall.
37. Jefferson and Clinton Sts., 11th Ward.
41. Willets Manufacturing Co., Brunswick Ave.
42. Crescent Pottery Co., Allen St. and Hankinson's Alley.
43. Greenwood Pottery Co., Cana' St.
51. Trenton Terra-Cotta Co., Third and Davies Sts.
52. Hamilton Rubber Co., 8th Ward.
53. City Railway Stables, Second and Lalor Sts.
54. New Jersey State Lunatic Asylum.
56. Liberty Engine House.
61. Woolverton Ave. and Cass St.
62.
63.

PHILIP FREUDENMACHER, Chief Engineer.

CHAS. C. DRAKE,
Sup't Fire Alarm Telegraph.

June 10th, 1888.

One of the earliest methods of sounding the alarm when fire erupted in Trenton was the ringing of bells, like in 1814, when a reward of $1 was offered to the first person to sound any of the Trenton Academy, First Presbyterian Church, or state house bells; whoever rang the second and third bells each received 50¢. By 1848, a system had been worked out whereby the number of rings from the city hall bell would signify to firefighters in which direction a fire was located. A fire alarm telegraph system was debated for many years. Finally, in December 1878, a private alarm connection was made between the Eagle and Harmony firehouses. The Gamewell company started running wires in Trenton for a trial in April 1879. The Utica Fire Alarm Telegraph Company then began installing its own hardware. The two systems were simultaneously tested for several months before the city, in March 1880, awarded a contract to Utica. But Utica's system was plagued by so many problems that common council reversed its decision that October and endorsed Gamewell. This 1888 poster shows the street locations of all the Gamewell alarm boxes then in use. (TFPL.)

Trenton's first ordinance regulating fire department operations was passed by common council on May 5, 1846. That ordinance, in part, established the office of chief engineer and laid out the rules by which the chief and assistants would be elected annually by delegates of the various companies. Over time, many changes were made to the election process. To provide better representation to firefighters in the southern wards, it was eventually agreed that one assistant chief would be elected from companies north of Assanpink Creek and another from those to the creek's south. Trenton's first volunteer chief engineer was Imlah Moore. He was followed, in order, by John P. Kennedy, John G. Gummere, Charles Moore, William J. Idell, Charles Bechtel, John McKechney, John B. Creed, Alfred S. Livingston, Samuel P. Parham, Jonathan S. Fish, John A. Weart, Levi J. Bibbins, Charles C. Yard, Thomas E. Boyd, William Ossenberg, Edwin S. Mitchell, Charles A. Fuhrman, Thomas Saxton, and Philip Freudenmacher. The drawing at left is of Weart, who was chief for several terms between 1866 and 1872. The photograph below shows the last volunteer leaders (1888–1892): from left to right, Assistant Chief (South) Frank Mangen, Assistant Chief (North) Walter Lanning, and Chief Philip Freudenmacher. (Both, MHFM.)

Philip Freudenmacher, an Eagle member since 1877, was elected to his first two-year term as chief engineer on April 10, 1888. Re-elected two years later, he remained in office until the volunteer department was disbanded. Freudenmacher was invited to stay on to lead the new paid department. Flattered by the fire commissioners' confidence in him but already committed to his responsibilities as local agent for a Philadelphia brewing company, he agreed only to serve a ceremonial 24-hour term—midnight to midnight, April 5, 1892—as Trenton's first paid fire chief. He later served as a fire commissioner and city councilman and then as Mercer County's sheriff. Freudenmacher died on Christmas night 1915, aged just 59. (MHFM.)

During his final year as volunteer assistant chief (1891–1892), Walter Lanning was paid by the city a stipend of $400 (Freudenmacher, as chief, received $600). Lanning started his firefighting career in 1880 with Harmony Fire Company. He joined the paid department in late 1892 as a hoseman, then was promoted to captain of Engine 5 on January 5, 1893, and became assistant chief on February 27, 1901. He retired on January 10, 1921, and died at the age of 79 on May 25, 1931. (MHFM.)

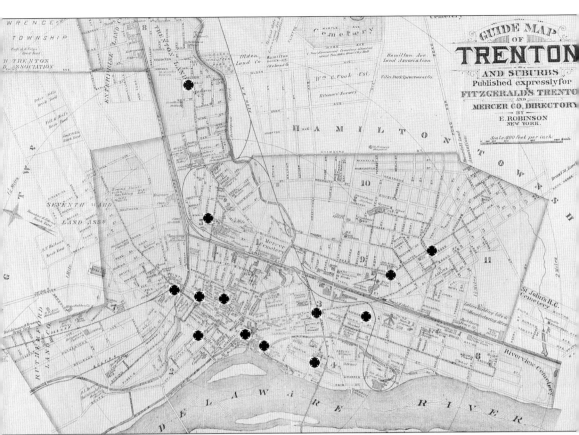

Crosses on this 1889 map mark the firehouse locations of the 13 companies that made up Trenton's volunteer department in its final years: America Hose and Engine Company at 522 Perry Street, opposite Ewing Street, east of the Delaware and Raritan Canal; Delaware Engine Company on South Warren Street, between Fall and Bridge Streets; Eagle Engine Company on South Broad Street, near Centre Street; Good Will Engine Company at 224 South Warren Street, between Washington (now Lafayette) Street and Assanpink Creek; Hand-In-Hand Engine Company at 29 North Willow Street, between West Hanover and State Streets; Harmony Engine Company at 317 North Warren Street, near Tucker Street; Liberty Engine Company, the northernmost, on North Clinton Avenue, near Olden Avenue; Lincoln Hook and Ladder Company, the farthest east, at 1005 South Clinton Avenue; Mutual Engine Company at the corner of South Clinton Avenue and Dye Street; Ossenberg Hose Company at Second and Furman Streets; Trenton Hose Company on East Hanover Street, near Greene (now Broad) Street; Union Engine Company at 16 Perry Street, between Warren and Greene Streets; and Washington Hook and Ladder Company at 333 South Warren Street, between Factory Street and the Trenton Water Power Canal. (TFPL.)

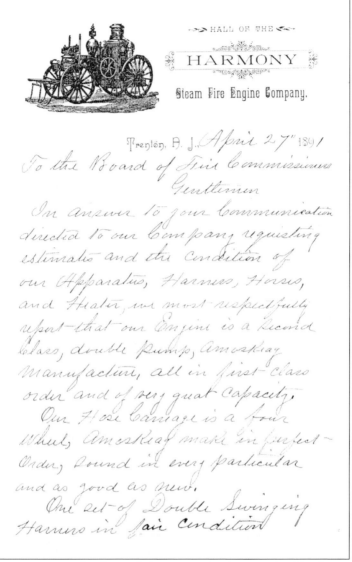

In April 1891, in preparation for the paid department's takeover, the 13 volunteer companies each provided city fire commissioners with written estimates of the value of their apparatus and other equipment. Shown here is the first page of Harmony Fire Company's letter valuating its assets—including its Amoskeag steamer, four-wheeled Amoskeag hose carriage, three horses, harnesses, and heater—at $3,000. The companies' various estimates totaled over $40,000. The commissioners met with the companies' representatives on February 27, 1892, and after some haggling agreed to pay less than $35,000 to purchase all the apparatus and equipment. Harmony's asking price of $3,000 was agreed to. The America's engine and the Washington's ladder truck were already owned by the city, as were all the companies' current firehouses, the construction of which the city had funded. After the paid department was in service, the Trenton Hose and Ossenberg Hose firehouses were sold, both buildings having been deemed unsuited to the department's future needs. Over the next few decades, new firehouses were built for Engine 5 (1898), Engine 1 and Truck 1 (1902), Engine 3 (1903), Engine 2 (1914), and several newly formed companies (Engines 7, 8, and 9). (MHFM.)

Shown here is the grand parade held at 3:00 p.m. on Monday, April 4, 1892, to bid farewell to Trenton's volunteer fire companies. Over 600 firefighters, both active and retired, marched accompanied by several hundred more dignitaries, police officers, and band members. The fire engines were so well polished they "shone as brightly as ever," the *Trenton Times* reported. A reviewing stand was set up outside city hall, and homes and businesses along the route were adorned with flags and bunting. Many businesses respectfully closed early, and thousands of spectators lined the streets. That evening, a banquet with seating for 650 was held at Turner Hall. Over $3,000 donated by appreciative Trentonians paid for the feast, which included roast turkey, cranberries, fried oysters, veal, ham, ice cream, cake, and more. Mayor Daniel J. Bechtel and others later gave speeches. While the volunteers' last actual alarm was a small blaze on North Montgomery Street on March 27, 1892, the last major fire they fought was the one that destroyed the insulated wire department braiding mill of the John A. Roebling's Sons Company off South Clinton Avenue, causing an estimated $250,000 in damage and injuring several firefighters, on February 4, 1892. (Both, TFPL.)

Two

1892–1920

Following a year and a half of planning and legal wrangling, fire commissioners on March 16, 1892, chose from a large pool of candidates the men who would become Trenton's first fulltime paid firefighters. Experienced members of the soon-to-be-replaced volunteer companies made up the majority of those selected for the various uniformed and civilian positions—one chief engineer, two assistant chief engineers, eight captains, six company engineers, six stokers, eight drivers, two tillermen, 24 hosemen, 10 laddermen, one fire alarm telegraph superintendent, one alarm lineman, one clerk, and two janitors. Annual salaries for the new hires ranged from $1,200 for the chief and $800 apiece for the assistant chiefs to $750 for captains and $600 for hosemen and laddermen.

The commissioners also decided which firehouses the paid department would use for day-to-day operations: Union (Fire Headquarters), Good Will (Engine 1), Eagle (Engine 2), Mutual (Engine 3), Liberty (Engine 4), Harmony (Engine 5), America (Engine 6), Washington (Truck 1), and Lincoln (Truck 2). The Delaware and Hand-In-Hand were designated storage and repair facilities (although Hand-In-Hand's North Willow Street building briefly housed Engine 6 until renovations were completed to the America's building, then in 1894 Truck 1 was relocated there and remained until moving, with Engine 1, into a newly built firehouse on West Hanover Street in 1902).

As April 4, 1892, drew to a close—the volunteers' farewell parade and testimonial dinner mere hours behind them—Philip Freudenmacher and the commissioners gathered. At midnight, as April 5 began, the man who would forever be known as Trenton's last volunteer chief and its first paid chief rang out the old and rang in the new by transmitting over the fire alarm telegraph three taps that symbolically announced the start of the paid department. He and the commissioners then made a tour of all the city's firehouses. Twenty-four hours later, again at midnight, Freudenmacher stepped down and turned leadership of the department over to his successor, William McGill.

This period, not surprisingly, was marked by great change. As Trenton's population increased—going from about 60,000 in 1892 to nearly double that by 1920—and the city turned into a major industrial center, the department grew with it. New companies, such as Chemical 1 and the Flying Squad auxiliary engine, were formed, and several new firehouses were built. The transition to motorized apparatus began in 1907, and in 1919, the last fire horses were retired from service.

William McGill was an experienced firefighter prior to becoming paid chief on April 6, 1892. Long affiliated with the America, he served as that company's treasurer from 1870 until at least 1896 (the old companies continued holding meetings and annually electing officers long after the volunteer department's disbanding). As a volunteer, McGill served as an assistant chief in 1873–1874 and was five times (1878 through 1886) unsuccessfully nominated for election as chief engineer. While responding to a general-alarm blaze on September 24, 1895, McGill was thrown from his chief's wagon when his horse stumbled. The wagon overturned, landing on McGill's leg. Bruised but not badly hurt, McGill freed himself, righted the wagon, and continued to the fire. A year later, he collapsed from exhaustion at another general alarm. Hoseman Charles Wood was killed at that blaze. The line-of-duty death of Wood (who had also been an America volunteer) greatly affected McGill, as did the death in 1900 of his 26-year-old son Samuel. McGill's health deteriorated, leaving him unable to work for months at a time. He died aged 61 on January 25, 1901. Mourners at his funeral included fire chiefs from Philadelphia, Paterson, and Lambertville. (Left, TFPL; below, MHFM.)

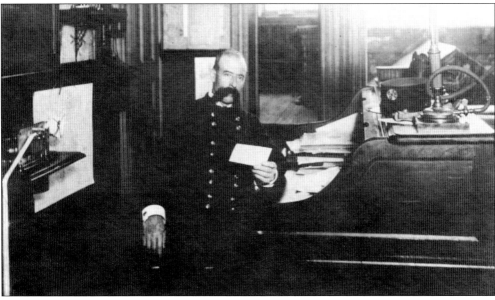

FIRE ALARM TELEGRAPH STATIONS WITH ASSIGNMENTS AND INSTRUCTIONS.

Box	Stations	First Alarm — Districts Covered By				Second Alarm — All Districts Covered By			
		Engines	Truck	Engines	Truck	Engines	Truck	Engines	Truck
3	City Hall	1, 2, 5	1	6, 4	2	6, 4	...	3	2
4	E. State and Canal Sts	1, 4, 6	1	2, 5	2	2, 5	...	3	2
5	Fire Headquarters, Perry and Broad	1, 5, 6	1	2, 4	2	2, 4	...	3	2
6	Willow St. and Bel. Del. R. R.	1, 5	1	6, 2	2	6, 2	...	3, 4	2
7	Factory and S. Broad Sts	1, 2, 5	1	6, 3	2	6, 3	...	4	2
8	Prospect and Bellevue Ave	1, 5	1	6, 2	2	6, 2	...	3, 4	2
12	New Jersey Steel and Iron Co	1, 2, 3	1	5, 6	2	5, 6	...	4	2
13	Fall and Warren Sts	1, 2	1	3, 5	2	3, 5	...	4, 6	2
14	Broad near Centre St	2, 3	2	1, 5	1	1, 5	...	4, 6	1
15	Furman and Second Sts	2, 3	2	1, 5	1	1, 5	...	4, 6	1
16	Market and Jackson Sts	1, 2	1	3, 6	2	3, 6	...	4, 5	2
17	Front and Stockton Sts	1, 2	1	6, 5	2	6, 5	...	3, 4	2
21	Warren and Lafayette Sts	1, 5	1	6, 2	2	6, 2	...	3, 4	2
23	State and Willow Sts	1, 5	1	6, 2	2	6, 2	...	3, 4	2
24	Pennington Ave. and Warren St	1, 5	1	6, 2	2	6 2	...	3, 4	2
25	Perry and Ewing Sts	1, 4, 6	1	5, 2	2	5, 2	...	3	2
26	Clinton and Perrine Aves	1, 4, 6	1	5, 3	2	5, 3	...	3	2
27	Brunswick Ave. and Rose St	5, 6	1	1, 4	2	1, 4	...	3, 2	2
28	Clinton and Monmouth Sts	4, 6	2	1, 5	1	1, 5	...	3, 2	1
31	S. Clinton and Dye Sts	2, 3, 6	2	1, 5	1	1, 5	...	4	1
32	Centre and Landing Sts	2, 3	2	1, 5	1	1, 5	...	4, 6	1
34	Clinton and Hamilton Aves	2, 3	2	1, 6	1	1, 6	...	4, 5	1
35	Chestnut Ave. and Bayard St	2, 3	2	1, 6	1	1, 6	...	4, 5	1
36	Police Station, S. Broad St	2, 3	2	1, 6	1	1, 6	...	4, 5	1
37	S. Clinton and Liberty Sts	2, 3	2	1, 6	1	1, 6	...	4, 5	1
41	Willets Pottery, Brunswick Ave	4, 5, 6	1	1, 2	2	1, 2	...	3	2
42	Crescent Pottery, Allen St	1, 5, 6	1	2, 4	2	2, 4	...	3	2
43	Greenwood Pottery Co	2, 4, 6	1	1, 3	2	1, 3	...	5	2
45	E. Hanover and Stockton Sts	1, 6	1	2, 5	2	2, 5	...	4, 3	2
51	Trenton Terra Cotta Co	2, 3	2	1, 6	1	1, 6	...	4, 5	1
52	Hamilton Rubber Co	4, 5, 6	2	1, 3	1	1, 3	...	2	1
53	Centre and Lalor Sts	2, 3	2	1, 6	1	1, 6	...	4, 5	1
54	New Jersey State Lunatic Asylum	1, 5	1	2, 6	2	2, 6	...	3, 4	2
56	Clinton and Olden Aves	4, 6	2	3, 5	1	3, 5	...	1, 2	1
61	Cass and Woolverton Ave	2, 3	2	1, 6	1	1, 6	...	4, 5	1
62	Trenton Watch Co	2, 3	2	1, 6	1	1, 6	...	4, 5	1
63	Clinton and Greenwood Aves	2, 6	2	3, 4	1	3, 4	...	1, 5	1
64	St. Francis Hospital	2, 3	2	1, 6	1	1, 6	...	4, 5	1
72	Warren and W. Hanover Sts	1, 5, 6	1	2, 4	2	2, 4	...	3	2
73	Central Police Station	1, 5	1	2, 6	2	2, 6	...	3, 4	2
81	Prospect and W. State Sts	1, 5	1	2, 6	2	2, 6	...	3, 4	2
112	Mill and Warren Sts	1, 2, 5	1	3, 6	2	3, 6	...	4	2
113	Mulberry St. and Del. & Rar. Canal	3, 4, 6	2	2, 5	1	2, 5	...	1	1
114	Trenton Iron Co	1, 2, 3	2	5, 6	1	5, 6	...	4	1
115	Murray & Whitehead	3, 4, 6	2	2, 5	1	2, 5	...	1	1
116	Lalor and S. Canal Sts	1, 2, 3	2	5, 6	1	5, 6	...	4	1
117	A. Exton & Co	1, 2, 3	2	5, 6	1	5, 6	...	4	1
123	Calhoun and Spring Sts	1, 5	1	2, 6	2	2 6	...	3, 4	2
124	Reservoir St. and Pennington Ave	1, 5	1	2, 6	2	2, 6	...	3, 4	2
127	City Hospital	4, 5	1	1, 6	2	1, 6	...	2, 3	2
135	Clinton and Chestnut Aves	2, 3	2	1, 6	1	1, 6	...	4, 5	1
137	Cummings and Anderson Sts	2, 3	2	1, 6	1	1, 6	...	4, 5	1
162	S. Broad Street Station	1, 2, 3	2	5, 6	1	5, 6	...	4	1
223	Calhoun St. and Feeder Bridge	1, 5	1	2, 6	2	2, 6	...	3, 4	2
212	E. State and Monmouth Sts	3, 4, 6	2	1, 2	1	1, 2	...	5	1
241	Centre and Cass Sts	2, 3	2	1, 6	1	1, 6	...	4, 5	1

This chart, taken from William McGill's first annual report (for the year ending February 28, 1893) as chief of the paid department, lists the locations of every alarm box in Trenton's Gamewell fire alarm telegraph system at that time. Company assignments, by alarm, are also shown for each box. Two engines and one truck were assigned to every first alarm. A third engine was added to firsts sounded from boxes in high-hazard areas. Two more engine companies rolled when a second alarm was transmitted. All remaining companies responded when a third, or general, alarm was struck for any box. The paid department's first alarm was received at 10:10 p.m. on April 7, 1892, for a small fire on East State Street. Another fire on April 28, 1892, at Brunswick and Princeton Avenues prompted McGill to order his first general alarm, but first-due companies succeeded in quickly controlling the blaze, and the extra help proved unnecessary. The next time McGill transmitted a general alarm, however, all hands went to work. That May 12, 1892, inferno at Perry and Carroll Streets completely engulfed Maddock and Sons' massive five-story pottery building. (MHFM.)

Engine 2 members are shown outside their original quarters, the old Eagle firehouse at 417 South Broad Street, near Centre Street. The photograph above was taken not long after the paid department took over, as both the building and the Clapp and Jones steamer still bear the number 3 previously used by the Eagle. Next to the engine's rear wheel is Capt. Harry Pennington, Engine 3's skipper from 1892 until his transfer in 1914 to command newly created Engine 8. The image below is from sometime after Engine 2 received one of the three new American LaFrance Metropolitan steamers delivered in July 1904. In December 1914, Engine 2 moved to its new firehouse at 503–505 South Broad Street, near Bridge Street. The old Eagle firehouse then became a department repair facility. The work schedule of Trenton's first paid firefighters was a difficult one. Men were on continuous duty, working 24 hours a day, seven days a week, like the military. Except for three one-hour meal periods per day (allowing those living close to visit their families) and 24 hours of leave every eighth day, the men literally lived in their respective firehouses and answered every alarm. (Both, MHFM.)

Engine 3 members pose outside their quarters (Mutual's old firehouse) around 1893–1894. From left to right are Hoseman William Temple (on hose carriage), Capt. Harry Braker, Hoseman John Henry, Hoseman Charles Ritter, Hoseman Louis Scheidnagel, Driver Frank Schollenberger (on engine), Stoker Thomas Tighe, and Engineer Rostene Ayres. This building (South Clinton Avenue and Dye Street) was demolished in late 1901 to make way for expansion of the John A. Roebling's Sons Company. On October 14, 1901, Engine 3 relocated to Truck 2's South Clinton Avenue quarters, and remained there until moving into its new South Broad Street firehouse on August 21, 1903. (MHFM.)

From left to right, Engine 5's John Phillips, John Kane, and Samuel Watson pass the time between alarms around 1905. At first, firefighters received no vacation. After successfully petitioning the city, the men were each finally granted one week of annual summer vacation starting in 1893. By 1909, annual vacation leave had grown to 12 days apiece, and later that year, their regular 24-hour leave was expanded to every sixth day instead of every eighth. A significant increase in off-duty time came in 1921 with the hiring of over 70 new men and the creation of a second platoon. (MHFM.)

The first home of Engine 5 was the old Harmony firehouse (317 North Warren Street). It was built in 1861, a year after the city agreed to construct new quarters for Harmony and Hand-In-Hand. That the facade still bears Harmony's name and founding date suggests the image above is from soon after the paid department started. Based on accompanying identification of the men, it was probably taken around 1895. From left to right are Hoseman Samuel Watson, Hoseman Fred Matheson, unidentified (possibly Hoseman Edward Tracy), Capt. Walter Lanning, Driver John Kane, Engineer Abram Chamberlain, Stoker Abram Chapman, and Hoseman George Howard. Finding the building not suitable to long-term needs, fire commissioners decided in 1896 to build a new firehouse for Engine 5. A location at Brunswick Avenue and North Alley was at first chosen, but this was abandoned, and property was acquired at Pennington Avenue and Willow Street. Construction started in September 1897, and Engine 5 moved into its new home the following April. The c. 1895 photograph below was taken outside Fire Headquarters. From left to right are Chamberlain, Watson, Howard, Assistant Chief Charles Allen, Chief William McGill, Assistant Chief James Bennett, Lanning, Chapman, Matheson, and Kane. (Both, MHFM.)

Lincoln Hook and Ladder Company's wood-frame firehouse at 1005 South Clinton Avenue was the first home of Truck 2. The April 18, 1892, *Trenton Times* described the building as "cramped," measuring just 14 feet wide by 67 feet long, with a ceiling only 8 feet high in parts. In July 1892, Truck 2 was temporarily relocated to the old Hand-In-Hand firehouse on North Willow Street, and the old Lincoln structure razed. A new firehouse, seen here, was quickly constructed in its place. Truck 2 moved into its new home on December 24, 1892. The c. 1895 photograph above shows, from left to right, Capt. Richard S. Fearnley, Driver Charles H. Tindall, Tillerman August H. Mundt, and Laddermen William Stackhouse, Joseph Hubbs, William Cubberley, Samuel McGill (brother of Chief William McGill), and Stephen McClarey. The image below is likely from April 1896, when Truck 2's Leverich apparatus was temporarily reassigned to Truck 1 while the rig was being repaired following an accident. Truck 2, meanwhile, used a wagon carrying a variety of ground ladders. During 1901–1903, while Engine 3 shared Truck 2's quarters, the Leverich apparatus was housed in a temporary addition built onto the side of the firehouse. (Both, MFHM.)

Organized on January 11, 1888, the Trenton Exempt Firemen's Association was, for the most part, a social organization made up of veteran members of the volunteer companies. The association's membership swelled after the volunteer department disbanded. The old Washington Hook and Ladder firehouse on South Warren Street (which Truck 1 used until moving in July 1894 to the old Hand-In-Hand building) was offered to the Exempts in 1895 and quickly became a repository for relics and memorabilia of the volunteer era. About that time, members of the old Good Will decided to locate their old hand engine for the collection. After finding their former apparatus irreparably dismantled in a junk shop, they searched for another rig similar to it. In 1898, they purchased for $100 an old hand engine from Manheim, Pennsylvania. This apparatus, which Manheim's Hope Fire Company had acquired in 1847, had originally been built around 1832 for Philadelphia's Globe Fire Company. Good Will members then paid to have the rig fully restored at the Trenton carriage factory of Valentine and Weeden, complete with murals painted by artist James Weaver. The finished engine, seen here, was presented to the Exempts on October 1, 1898. (Both, MFHM.)

For decades after they were replaced by the paid department, the volunteer companies continued to hold meetings and annually elect officers from among surviving members. While they no longer battled blazes, these old firefighters continued to participate in parades and trade war stories in the Exempts' firehouse. Shown here in his Exempt Firemen's Association uniform is John H. Carlile, an Eagle member who died in 1929. One of the last known volunteer meetings took place on February 7, 1940, when the Union's last three survivors gathered to mark the company's 193rd anniversary. The last of these men, Robert A. Ford, died in 1946. (MHFM.)

Shown here in 1932 is an old Trenton Hose Company carriage in the Exempts' quarters, 333 South Warren Street. This rig was sold to Princeton's Mercer Engine Company in May 1871, a few years after a new Trenton Hose carriage (page 19) went into service. The apparatus was bought back from Princeton in 1911 and paraded by Trenton Exempts in Norristown, Pennsylvania, in 1912. With few volunteers still alive, the Exempts' collection was broken up and the firehouse sold off by the city in 1941. The building was later damaged by a two-alarm fire on November 3, 1948. (MHFM.)

Formed from part of Hamilton Township, Wilbur Borough was created on April 24, 1891. For its first few years, Wilbur relied on Trenton for fire protection. In 1895, the borough formed its own volunteer fire department and spent $2,200 to buy the chemical engine seen here, a Babcock model equipped with two 60-gallon tanks. The apparatus's first run was for a small fire at Thomas Smires's grocery store on Walnut Avenue on July 4, 1895. Wilbur's volunteers had quarters in borough hall on South Olden Avenue (which in 1913 was renovated to house Trenton's new Truck 3). (MHFM.)

After Wilbur was annexed by Trenton on February 28, 1898, Chief William McGill created a new company, Chemical 1, to use the old borough apparatus. Wilbur's former volunteer driver, William J. Barber, was hired as Chemical 1's driver. Barber was later promoted to lieutenant in 1907, then captain in 1913. On April 11, 1917, Captain Barber was driving to a fire in Hamilton Township when Chemical 1's motorized apparatus slid into a South Broad Street canal bridge. Assistant Chief Walter Lanning was thrown through the windshield and injured, as were Barber and two others. Barber retired in 1928 and died in 1940. (MHFM.)

Charles S. Allen spent over a decade battling blazes as a Hand-In-Hand volunteer before his hiring as first assistant chief in 1892. He served as acting chief for much of 1900 while William McGill was incapacitated and, following McGill's death, was formally promoted chief on February 1, 1901. Like McGill, Allen was not afraid to share the risks faced by his men. During a December 24, 1895, fire in a furniture factory, Allen lost control of a hoseline and fell head-first from a ladder. He was knocked unconscious and suffered multiple injuries. He was injured again on February 3, 1907, when a wall collapsed during a general-alarm fire at Imperial Porcelain. On October 2, 1907, he fell and broke a rib after flooring burned away during a fire on Muirhead Avenue. Allen advocated for improvements to Trenton's water mains and purchased the department's first motorized apparatus. Ill health forced Allen to retire on August 19, 1911. He died three months later at the age of 55. (MHFM.)

Driving Truck 1 to a fire is Edward H. Hizer. Standing next to him is Capt. Charles A. Knoblauch. Alexander Grugan is the tillerman. The photograph was taken by Bernard Schnell outside his Princeton Avenue barbershop at some time prior to Hizer's 1907 line-of-duty death. (TFPL.)

The large four-story building constructed in 1884 at North Warren and West Hanover Streets by brothers Frank and Peter Clark was the scene of a major blaze that gutted the basement storehouses of the brothers' china and glassware store and two other businesses on April 29, 1885. Nineteen years later, on May 11, 1904, another fire, seen here, broke out in the same building, which was still owned by the Clarks but by then occupied by the Brand and Smith furniture store, operated by David H. Brand. A muffled explosion was reportedly heard by employees, and moments later, heavy smoke billowed up through the elevator shaft from the basement. Fueled by all the furnishings and barrels of furniture oil stored in the basement, flames rapidly spread via the elevator shaft to the floors above. With seamstress Maggie Mitchell in his arms, deliveryman Harry Voorhees jumped from a fourth-floor window to a neighboring shed roof to escape the firestorm. (Both, TFPL.)

Responding with Engine 5 and Chemical 1 on the first alarm, sounded at 2:58 p.m. on May 11, 1904, firefighters from Engine 1 and Truck 1 left their nearby quarters on West Hanover Street to see dense smoke pouring from almost every window and fire showing from the top of the Brand and Smith store. Assistant Chief James Bennett ordered a second alarm when he arrived, soon followed by a general alarm for all remaining companies. It took firefighters two hours to control the blaze. A year later, on July 2, 1905, a fire of suspicious origin damaged the David H. Brand and Company department store at East State and Montgomery Streets. In 1906, after a year-long investigation, authorities charged David Brand and his brother John with aiding and abetting the arson of their store. They were both found guilty following court testimony that—in stretching from November 26 through January 2, 1907—was then called the longest trial in Mercer County history. Among those appearing at the trial was Chemical 1 captain John Stiefbold, who testified that after extinguishing the smoky blaze, firefighters found that large amounts of a flammable liquid had been poured all over the stock, shelves, and floors of the business. (TFPL.)

This photograph was taken some time after Engine 1 and Truck 1 moved into their new firehouse at 51–53 West Hanover Street on August 7, 1902. The unusual apparatus shown is the department's water tower. Chief Charles Allen and fire commissioners Louis Diehl and William Ossenberg paid $225 for the Greenleaf/Babcock 45-foot tower at an auction of ex–Fire Department of New York equipment on May 27, 1901. The rarely used water tower was scrapped in the 1920s. On December 16, 1903, as the entire department was busy battling a general alarm at a West End Avenue factory, a boiler explosion occurred in the West Hanover Street firehouse. Engine 2 and Chemical 1 had to be released to extinguish the burning basement of the firehouse. (MHFM.)

While the city committed in 1904 to erecting a new firehouse in the area of Hamilton and Chestnut Avenues, it was not until May 27, 1907, that newly formed Engine 7 moved into its just-completed new home at Hamilton Avenue and North Anderson Street. Engine 7's first commander was Capt. John Stiefbold, and its first rig was a new American LaFrance Metropolitan steamer capable of pumping 800 gallons per minute that arrived in Trenton on September 30, 1906. The harnesses for Engine 7's steamer and hose wagon horses are visible in this undated photograph. (TFPL.)

The crew of Engine 3 in 1908 are, from left to right, Capt. Louis Scheidnagel, Engineer Rostene Ayres, Hoseman Frederick Henry, Stoker Thomas Tighe, Hoseman Christian Walter, Driver Samuel Homer, Hoseman Leon Carson, and Hoseman Charles Ritter. In use by Engine 3 at this time was one of the American LaFrance Metropolitan steamers delivered in 1904. (TFPL.)

Members of Engine 5, also in 1908, are, from left to right (first row) Lt. Samuel Watson, Capt. Jeremiah McGill, and Engineer John W. Phillips; (second row) Hoseman Anthony Moran, Hoseman John J. Farrell, Stoker Fred Van Horn, Hoseman John Donohue, and Driver Theodore Smith. Son of the late Chief William McGill, "Jerry" McGill rose through the ranks and became chief of the department himself in 1921. Engine 5's apparatus in 1908 was the 1887 Clapp and Jones steamer originally built for the Eagle volunteers and first used by Engine 2. (MHFM.)

Harry Braker, who had a decade of firefighting experience as a volunteer with the Union before he joined the paid department, was originally hired as a hoseman but was promoted to captain of Engine 3 in August 1892. An avid checkers player, Braker defeated South Clinton Avenue barber Anthony Marotte to be crowned "Checkers Champion of Chambersburg" in 1906. He remained Engine 3's skipper until severe asthma forced him to retire in September 1907. He died in 1940 at 81. (MHFM.)

James W. Bennett began his firefighting career in 1880 as a volunteer with Trenton Hose Company. Originally hired in 1892 as second assistant chief under Chief William McGill, he was named acting chief after McGill's successor, Charles S. Allen, retired on August 19, 1911. His promotion as the fourth chief of the paid department was made official on October 23, 1911. Bennett had several close calls during his career but perhaps none more so than when he was almost fatally electrocuted by an overhead electrical wire that fell on him during a December 5, 1907, fire at Billingham Iron and Machine Works on William Street. "The shock was terrible," he said in an interview later published by *Fire and Water Engineering*. Bennett retired on January 10, 1921. He was 72 when he died on February 23, 1927. (MHFM.)

A 10-horsepower steam-propelled Stanley runabout purchased in 1907 for Chief Charles S. Allen was the department's first motorized vehicle. Although very small by today's standards, the vehicle was still larger than most of its kind, as the department had it built with a second row of seating to accommodate a total of four men. Delivered to Trenton in Stanley's signature red color, the car was immediately sent to local carriage builder Valentine and Weeden's factory to be painted white, as the Stanley company had refused to do so. The car remained in service, used by Chief Allen and his successor, Chief James W. Bennett, from May 4, 1907, until May 1912, when it was reallocated for use by the superintendent of the city's crematory. It was replaced by the 1912 Stanley model seen here, a 20-horsepower five-seater. Designed for use as an emergency ambulance if needed, this new chief's vehicle was equipped with a collapsible stretcher and first aid supplies. From left to right are driver John L. Westenburger, Assistant Chief Walter Lanning, and superintendent of machinery Rostene Ayres. The photograph, taken in front of Engine 1 and Truck 1's West Hanover Street quarters, is undated but was likely taken in 1912, since Ayres died in March 1913. (MHFM.)

This was the department's second motorized apparatus, a Stanley steam-propelled chemical engine. After arriving in Trenton by special freight train from Newton, Massachusetts, on July 4, 1908, two Babcock 30-gallon copper tanks were installed under the supervision of Rostene Ayres, the department's newly appointed superintendent of machinery. These tanks contained sodium bicarbonate (baking soda), sulfuric acid, and water, which when mixed created an effective fire extinguishing agent. The new apparatus was also outfitted with two 3-gallon soda acid hand extinguishers, 400 feet of chemical hose, 500 feet of standard 2½-inch fire hose, axes, and rope. It was painted cream, with carmine-colored wheels. The lettering was in gold, shaded with blue. The new Chemical 1—which cost $3,300—responded to its first alarm on July 28, 1908, quickly extinguishing a small fire at 394 South Warren Street. On August 8, 1908, the new rig proved its 30-horsepower speed when, responding from Fire Headquarters on Perry Street, it beat first-due Engine 2 to a blaze at 92 South Clinton Avenue. This December 1908 image shows, from left to right, Hoseman Walter Allen, machinery superintendent Ayres, Hoseman George Voorhees, Lt. William Barber, and Capt. William Stackhouse. (MHFM.)

Engine #8, Trenton, N. J., June 1914

Shown here a few months after their newly formed company went in service on February 26, 1914, are Engine 8 members outside their quarters at Stuyvesant and Edgemere Avenues. Engine 8's first apparatus was this 1909 Webb previously used by Engine 1. On February 22, 1909, A.C. Webb, racecar driver and founder of Indiana-based Webb Motor Fire Apparatus Company, staged a demonstration in Trenton of one of his gasoline-powered rigs built using a Thomas Flyer chassis. In a competition starting from the West Hanover Street firehouse, the Webb apparatus had a stream flowing over the Trenton Battle Monument a full five minutes before Engine 1's 1904 American LaFrance horse-drawn steamer even started throwing any water. Several more demonstrations were made, including one in which the Webb sent water over the top of the First National Bank in Princeton just 18 minutes after leaving Fire Headquarters. Afterwards, city officials signed a $7,000 contract. This four-cylinder Webb—cream and carmine in color, with gold lettering shaded in blue—arrived in Trenton on August 28, 1909. Able to pump 600 gallons per minute, the new rig proved its worth when (as Engine 1) it responded during a blizzard to a fire at the Interstate Fairgrounds in Hamilton Township on December 26, 1909. While horse-drawn apparatus could barely navigate the snowy streets (Truck 2 went only two blocks before getting stuck), the Webb and Chemical 1 (operating its motorized 1908 Stanley) overcame snow drifts four feet tall to make the three-mile journey. While the grandstand was destroyed, other parts of the famous fairgrounds were saved. (MHFM.)

While the department's second Webb engine (in this undated photograph) also cost $7,000, it was an improvement over the first in that it was slightly larger and had a six-cylinder motor. The Webb company, having by then relocated to St. Louis, Missouri, was so busy in 1910 it could not supply to Trenton a second four-cylinder engine as contracted. Company president A.C. Webb happily offered the larger apparatus at no extra cost in recognition of the notoriety that came his way after Trenton, with its 1909 purchase, became the first department east of the Allegheny Mountains to operate one of his engines. After being exhibited at that year's international fire chiefs' convention in Syracuse, New York, the new rig arrived on August 31, 1910. Once in service, it was assigned to Engine 1. The 1909 Webb, in turn, was moved to headquarters for use by the new auxiliary engine company then being planned. While sometimes informally referred to as Engine 8 prior to that company's actual organization in 1914, the auxiliary had no official engine company number and was known simply as the "Flying Squad." The auxiliary was formed in early 1911 and staffed by the transfer of one man from each of the seven numbered engine companies then in service. The auxiliary went to all calls for service citywide, providing extra manpower as needed to first-due companies. The auxiliary remained numberless until it was officially renamed Engine 10 in March 1925. (MHFM.)

Shown here is the paid department's first headquarters. Located at 16 Perry Street, between North Broad and North Warren Streets, it was built by the city in the volunteer era for $5,285 after the old Union firehouse on the same site was torn down in October 1887. Union volunteers held a celebratory open house on April 5, 1888, but their time there was short, as the volunteer department's disbanding came just four years later. This firehouse served the paid department from April 5, 1892, until a newer, modern headquarters built just down Perry Street near Stockton Street opened on September 12, 1927. The old building was sold off and changed hands many times. It was last used as a church and was demolished in 1982 as part of a project to realign Perry Street. From left to right are the auxiliary company's brand-new 1913 Webb engine, Chemical 1's 1908 Stanley chemical engine (with Capt. Alexander Grugan in the officer's seat), and the department's 1912 Stanley chief's car (with Chief James Bennett in the rear). The auxiliary's new six-cylinder, 72-horsepower Webb arrived in Trenton in early July 1913, and this photograph was likely taken around then, because by July 20, Grugan had been reassigned to command the newly formed Truck 3. The auxiliary's old 1909 Webb was subsequently reassigned to Engine 8 when that company was created in 1914. On September 2, 1913, while responding to a fire, the 1913 Webb met with a freak accident when it ran into a horse on East Hanover Street. The horse, having had one of its legs torn completely off, had to be put down by a shot from a police officer's revolver. (MHFM.)

While not confirmed, these two photographs are believed to have been taken on April 24, 1913, when Engine 3's new engine was tested along the Water Power Canal behind the state house. The extra-sized first-class American LaFrance Metropolitan steamer, delivered from Elmira, New York, to Trenton a few days earlier, was guaranteed to pump 1,200 gallons per minute. During the testing, water was flowed through two lines of 3-inch hose and two lines of 2½-inch hose. A note on the back of one of the images identifies the man wearing the derby hat as J. Wallace Hoff, secretary of the city's board of fire commissioners. At first pulled by three horses, this apparatus—like most of the department's other steamers eventually were—was motorized by having a gasoline-electric Couple-Gear tractor attached to it in 1914. During a trial conducted in September 1914 after the install was completed, Engine 3 reached a speed of 21 miles per hour. Engine 2, also tractorized at the same time, crashed into a Jackson Street home during driver training on October 9, 1914. (Both, MHFM.)

Fire was a frequent occurrence at the Trenton properties of the John A. Roebling's Sons Company, wire-rope cable manufacturer and builder of the Brooklyn Bridge, among other accomplishments. Many of these were major blazes. During the volunteer era, a building used to make electric cables burned to the ground on July 20, 1891, and on February 4, 1892, flames destroyed the company's wire braiding mill. (Though not a company property, the John A. Roebling School, a city public school at Home Avenue and Beatty Street named in honor of the Roebling patriarch, was ravaged by fire on February 14, 1902.) Above, a general-alarm blaze on February 5, 1908, caused more than $250,000 in damages and destroyed several buildings, including Roebling's Elmer Street rope and carpenter shops. Another general-alarm fire on December 2, 1911, this time at the company's flat wire mill at South Clinton Avenue and Mott Street, resulted in losses totaling about $100,000. Below is yet another general-alarm blaze that swept through the Roebling stables on Swan Street on February 10, 1913. Although 15 tons of hay burned up, over 75 horses that had been inside were evacuated to safety. (Above, TFPL; below, MHFM.)

Shown here is the aftermath of another general-alarm blaze, which gutted the Roebling company's rod rolling mill on South Clinton Avenue on February 1, 1913. The fire started when a red-hot rod slipped off a roller into an oil pit beneath one of the machines, igniting an explosion. The building's 120 workers fled as "flaming oil was thrown in every direction by the swiftly revolving belt," the *True American* reported. Arriving to find the mill already a mass of flames, firefighters concentrated their efforts on saving the adjacent structures. Sixteen hoselines were put in service to control the blaze. Firefighters encountered many obstacles during the various Roebling fires. It was so cold at the 1913 stable blaze, for example, that several hosemen had their hands frozen to their nozzles. Oils stored throughout the flat wire mill fueled flames during the 1911 fire so quickly that the entire building was engulfed before first-due Engine 3 arrived. The 1908 blaze, meanwhile, took place on another bitterly cold day when frozen hydrants had to be thawed out and many firefighters suffered frostbite. The involved buildings were so large (one side of the two-story rope shop alone stretched nearly 600 feet) and filled with such highly flammable materials that flames were massive. (MHFM.)

The worst of the Roebling fires was, without question, the January 18, 1915, conflagration involving the company's Buckthorn Plant. To this day, it is still considered the largest blaze in Trenton history. The fire was believed to have been deliberately set, possibly by a German spy, as the company had just signed a contract with the British and French for war supplies. The plant, home of Roebling's insulated wire department, comprised several interconnected buildings (one five stories tall) on eight acres off Jersey and Tremont Streets. Sabotage to fire alarm wires reportedly gave the blaze a tremendous head start. Wind gusts fanned flames and carried away embers as big as baseballs. Radiant heat was intense. Fifteen nearby homes were also destroyed, and 20 others were damaged. For the first time, Trenton firefighters called for outside help. Engines and men from Princeton and Camden responded. Damage estimates ran up to $2 million. These two photographs, originally taken by Bain News Service, show the Buckthorn ruins. On November 11, 1915, another general-alarm fire of suspicious origin destroyed Roebling's new Elmer Street rope shop, resulting in $1 million more in losses. (Both, Library of Congress, Prints and Photographs Division; above, LC-DIG-GGBAIN-18230; below, LC-DIG-GGBAIN-18231.)

This photograph is believed to have been taken shortly after Truck 3 was organized in 1913. From left to right are Samuel Homer (driver), Capt. Alexander Grugan, Joseph Dydzulis, Edward Holland, Harry Applegate, James Cryan, Samuel Bramley, and Charles "Montana" Vandegrift (tillerman). The horse on the left was called George, and the other was Teddy. Truck 3 was put in service at 1:25 p.m. on July 20, 1913, and was quartered in the old Wilbur Borough Hall at South Olden and Walnut Avenues. The building, which was renovated to accommodate the truck, had previously been home to Wilbur's chemical engine. (MHFM.)

Following the June 8, 1913, arrival and subsequent testing of Truck 1's new Webb gasoline-electric apparatus (equipped with a 75-foot aerial ladder), Truck 1's old horse-drawn Dederick rig (which was delivered to Trenton on September 3, 1904, and was equipped with a 65-foot aerial) was reassigned to Truck 2. This, in turn, freed up Truck 2's old horse-drawn Hayes rig (which the city took delivery of in April 1898 and which had a 55-foot aerial) for use as Truck 3's first apparatus. Capt. Alexander Grugan and Truck 3's crew are seen here with that rig prior to an April 17, 1915, parade. (MHFM.)

Pictured here after the April 17, 1915, parade are Engine 5's 1913 motorized hose/chemical wagon and horse-drawn 1904 American LaFrance Metropolitan steamer. The wagon was assembled using a Webb chassis by local carriage-makers Fitzgibbon and Crisp following their successful build of a similar apparatus for Engine 3. From left to right are August Conway (driving), Lt. Samuel Watson, Walter Tettemer, Frederick Slover, Edward LaBaw, and John Farrell. With the steamer are Theodore Smith (left) and John Phillips. Two weeks later, many of these men served as pallbearers for Slover, who was fatally crushed on April 27 when a wall collapsed on him during a three-alarm fire. (TFPL.)

This horse-drawn 1881 Amoskeag steamer, still in use at the time by Engine 6, was among the rigs that were paraded on April 17, 1915. In fact, Engine 6 was the last Trenton company to use horses. Engine 6's team made its final run, to a small fire on Forst Alley, on August 26, 1919. Four days later, the company started using a motorized Thomas/Northern pumper. The horses, having been given to the police and sanitation departments, were gone from Engine 6 a week later. (MHFM.)

Engine 8's motorized 1909 Webb pumper is seen here lined up on East Hanover Street, where fire department units were inspected by Mayor Frederick W. Donnelly before the start of the April 17, 1915, parade. The procession featured city police officers, firefighters, and sanitation workers and was arranged by public safety director George B. LaBarre as a way to show off those departments' resources. Each rig in the parade bore signs listing its manufacture year and value. All nine engine companies (including the unnumbered auxiliary), all three truck companies, and Chemical 1 took part. Engine 9 was organized on September 10, 1918, and stationed in a newly built firehouse at 1100 Brunswick Avenue, while the auxiliary company subsequently became Engine 10. (MHFM.)

Shown in this undated photograph is Engine 3's motorized combination hose and chemical wagon. Using a chassis obtained from the Webb Motor Fire Apparatus Company, Fitzgibbon and Crisp built the body in their factory on Bank Street in Trenton in 1913. It was equipped with a Morse "turret pipe" (deluge gun) and 40-gallon chemical tank and carried 1,400 feet of 2½-inch fire hose, 400 feet of 1-inch chemical hose, ladders, two extinguishers, hand tools, and more. (MHFM.)

Three

1921–1950

During 1893, its first full calendar year of service, the paid department had a uniformed strength of 69. This included the messenger, a firefighter who (similar to the chief's aides of later decades) was assigned to headquarters but could fill in as needed when injury or illness left a company short-staffed. Also included was the extra engineer hired in May 1892 to maintain reserve apparatus. By 1922, following the phased implementation of the two-platoon work schedule a year earlier, the department had grown to 201 uniformed members. Under the new system, two equally divided platoons worked 10- and 14-hour shifts, with no time off for meals. The platoons alternated days and nights every three days. Annual vacations, based on rank, varied from 14 to 30 days.

In June 1927, in preparation for the transition three months later to the new headquarters at 244 Perry Street, the telephone switchboard, Gamewell alarm transmitter, and other alarm equipment were moved from the old headquarters at 16 Perry Street into the Trenton Electrical Bureau at 29 West Hanover Street. Also in June 1927, in the leadup to the new headquarters' opening, Engine 6 and Truck 3 moved firehouses. Truck 4 was organized in 1928, and in 1929, each Gamewell alarm box was given a new zoned four-digit number to make identifying its physical location far easier.

Built behind the new headquarters was a spacious repair facility for use by department mechanics (who, since Engine 2 moved out in 1914, had been working from the old Eagle firehouse on South Broad Street). A machine shop, blacksmithing facilities for welding and cutting, a paint shop, and storerooms were on the first floor, with a carpentry shop and other facilities on the second floor.

Other notable changes during this era were Chemical 1's disbanding in 1931, the reclassification in 1936 of assistant chiefs to deputy chiefs (and subsequent discontinuation of the assistant rank), and the elimination of the lieutenant and battalion chief ranks (first used by the department in 1906 and 1915, respectively) with promotion of the men holding those titles in 1943.

During World War II, hundreds of city residents volunteered for 22 weeks of training to join Trenton's Fire Reserve Corps. Dozens of these auxiliary firefighters helped out at a general-alarm blaze in a four-story North Broad Street food market on April 18, 1943. That year, the department found itself short 43 men due to retirements and 25 members being away on active military duty.

Shown here in 1924 by the hydrant are Francis Apgar Sr. (left) and Joseph O'Donnell. In the front seat of Chemical 1 (the remodeled Webb) are Herbert Lukens driving, followed by Capt. William J. Barber and Russell Schuchardt. In the back seat are, from left to right, Fritz F. Schmiedel, James Peters, and William Scales. Posing on and beside the auxiliary company's 1923 American LaFrance are, from left to right, Edward Holland Sr., Edward Kearns, John A. Dill, John Adams, Benjamin Durham, Clarence Muhs, James Irwin, Walter Treftz, Thomas Dovgala, Conrad Lenius, Lt. George Ecker, Anthony Felice, Horace Heck, Charles J. Rainbow, and John Bissett. In the windows are Battalion Chief Louis Scheidnagel (left) and Assistant Chief Roland Weigand. (MHFM.)

This 1924 image shows Engine 1's 1918 American LaFrance and Truck 1's 1913 Webb apparatus. From left to right are (first row) John Szagany, John Seaman, Norman Shotten, Andrew Tomko, Michael Maloney, Lt. Homer Collins, Capt. William J. Mitchell, Joseph Manning, William Buckley, Stephen Conrad, John Demonski, John Powers, unidentified, Raymond Fischer, Andrew Irwin, Michael J. Egan, and George Voorhees. The engine driver is Alvin Slack, and Henry Reitmeier is next to him. On the truck, from left to right, are Arthur Phillips (tillerman), Edward Frey, Edward R. Gore Sr., John Ewing, James Reams, Lt. Valentine J. Burkhauser, Angus McDougal, and Lt. Joseph Scudder. (MHFM.)

This photograph shows Engine 2 around 1915–1920. The department's 1921 annual report lists Engine 2's apparatus as a tractorized 1904 second-class American LaFrance Metropolitan steamer and a Kissel hose wagon equipped with a 35-gallon chemical tank. The Kissel wagon may be the same one later used by Engine 3 (page 70). The tractor was added to the steamer in September 1914. A few months after that, on January 6, 1915, two civilians were tragically killed when, while responding to a fire, Engine 2's tractor skidded on wet pavement along South Broad Street and the rig ran up the curb into a sidewalk full of pedestrians. (MHFM.)

This was how Engine 2, at South Broad and Bridge Streets, looked a few years later in 1924. The rig is one of three American LaFrance triple-combination pumpers ordered in 1922 and put in service after their arrival in early 1923. This apparatus, like many others in the department, carried a collapsible life net. From left to right are Lt. George R. Hundt, James Messeroll, John Brown, Anthony H. Werr, Paul Wiedenhofer, Thomas Wagner, William R. Dobbs, William J. Sampson, John Joyce, George Schweder, William Buckley, William M. Stackhouse, and Clement McCann. (MHFM.)

In 1903, Engine 3 moved into this brand-new firehouse at South Broad Street near Jennie Street (now Hudson Street). In 1924, as shown here, the company was using this 1915 Kissel hose/chemical wagon and tractorized 1913 extra-sized first-class American LaFrance Metropolitan steamer. From left to right are Capt. Thomas J. Phelan, August H. Bremer, Emerson Hill, Christian Walter, Paul Prykanowski, Charles Howard, William H. Coker, Michael Husosky, Patrick Kelly, Anthony Chato, Thomas Ricketti, Frederick C. Banks, Harry J. Frey, and Lt. Peter J. Concannon. (MHFM.)

Shown here in 1924 outside their North Clinton Avenue quarters are members of Engine 4. From left to right are (first row) Peter Palz, Matthew J. Maloney, Capt. Otto W. Yaeger, John Quinn, and Andrew J. Uhaze; (second row) James Dolan, William Boorer, William Tettemer, John T. Dempster Sr., Frank Sutts, George J. Blurton, George W. McClellan, and Albert S. Dura. The hose wagon is listed in department documents as being a 1915 Boyd equipped with a 35-gallon chemical tank. The tractor-drawn steamer is a 1917 first-class American LaFrance Metropolitan. (MHFM.)

Shown here at their Pennington Avenue firehouse, Engine 5 members in 1924 were assigned this tractorized 1904 second-class American LaFrance Metropolitan steamer and 1913 Webb hose and chemical wagon. From left to right are Theodore "Dory" Smith, Alonzo Pullen, Peter J. Schafer, Clarence Crosby, Earl S. Campbell, John J. Pfeiffer, John J. Farrell, John "Jake" Molnaur, George Anderson, John Brennan, William A. Dale Sr., Lt. Samuel C. Watson, and Capt. Walter S. Allen. (MHFM.)

In 1924, as shown here, Engine 6 was still operating from the old America firehouse at 522 Perry Street. Posing with their new 1923 American LaFrance triple-combination pumper are, from left to right, Capt. Augustus "Gus" Conway, Fred Dettmar, Edward Randall, James J. Kennedy, George Fako, Arthur Stinger, Owen McCarthy, Ferdinand C. Kruger, Joseph Schaller, Martin Dura, and Lt. Thomas P. Gilligan. In the driver's seat is John O'Donnell, and next to him is William J. Megules. On June 27, 1927, Engine 6 was permanently reassigned to the firehouse at South Olden and Walnut Avenues, and Truck 3 relocated from there to share Engine 9's quarters on Brunswick Avenue. (MHFM.)

Like all but one of the images reproduced on pages 68 through 76, this photograph of Engine 7 was taken by photographer J. Harry Kidd for the souvenir booklet given out at a May 23, 1924, dance held as a fundraiser for the city firefighters' union, Firemen's Mutual Benevolent Association Local 6. In 1924, Engine 7 used this 1906 second-class American LaFrance steamer and another rig described in department records as a 1918 Thomas/Northern "combination pumper, hose car, and chemical wagon." Engine 7's roster at this time included Capt. Frederick Toft, Lt. William Feist, John Cronin, George Groh, Albert Hallett, William Hartz, William Henry, Peter Howard, Clarence Leip, Simon McNamara, Andrew Sagedy, Edmund Schaller, William Stocker, and John Turner. (MHFM.)

For a time in 1924, Engine 8 used this 1918 Thomas/Northern combination apparatus. This rig was damaged on November 10, 1923, when, after losing a wheel, it struck a tree on Stuyvesant Avenue while going to a fire. Later, in 1924, the apparatus was taken out of service and Engine 8 was reassigned one of the old Webb pumpers. From left to right are Edward Sweeney, Walter Tettemer, Frank Hoffman, Lawrence Begley, Morris Moses, Stephen Tracey, Frank J. Jablonski, Martin Conry, John Bumbera, Joseph Bunsky, John Messler, Walter Martin, Lt. Samuel Homer, and Capt. Fred Foster. (MHFM.)

Engine 9 in 1924 also had a 1918 Thomas/Northern "combination pumper, hose car, and chemical wagon." Engine 9 members had this firehouse on Brunswick Avenue all to themselves from their organization in 1918 until Truck 3 was relocated here on June 27, 1927. While Truck 3 moved out in 1965, Engine 9 stayed on Brunswick Avenue until moving into a brand-new firehouse on West State Street in 2003. From left to right are Lt. Thomas Moran, Joseph Eagle, Thomas O'Hara, John H. Sawyer, Eldridge S. Waldron, William Nelson, Adolph F. Doldy, George Disbrow, Frank Jordon, George Woolverton, John Uhaze, John Entwistle, and Thomas Horrobin. (MHFM.)

Truck 2's apparatus in 1924 was this 1917 model built by the Couple-Gear Freight Wheel Company of Grand Rapids, Michigan. Equipped with a 65-foot aerial, the truck reportedly cost the city $11,000. In 1925, it was refurbished and a new American LaFrance tractor installed. Individual identification of the men shown is not available except for the rightmost two standing: Lt. George Weigand (left) and Capt. Harry Kline. Others assigned to Truck 2 in 1924 were John Bayen, Joseph Boytos, John Breza, James Convery, Victor Ecker, John Falkowski, Andrew Franko, Andrew Groffie, Joseph Landerkin, William Miller, Andrew Pavlicin, and Herman Schweder. (MHFM.)

A Thomas city service truck (first put in service on September 7, 1918) was Truck 3's apparatus in 1924. The straight-bodied rig carried 13 ground ladders, including one 50-footer and two pompiers, as well as a life net and two 3-gallon hand extinguishers. From left to right are (first row) Capt. James Hann, Lt. Charles Jenkins, Robert F. Van Hise, Harry Higginbottom, Paul F. Gaertner, Clarence Rust, David Burke, Charles Randolph, and Anthony P. Jablonski; (second row) George Malone, James Feehan, Michael Uhaze, Frank A. Pinto, and DeWitt Wolcott. (MHFM.)

Jeremiah McGill joined the department as a hoseman on May 31, 1899, while his father William McGill was still chief. "Jerry" was made an acting captain in 1905 and permanently promoted to captain on June 26, 1907. He was elevated to battalion chief on September 15, 1918, and then promoted to assistant chief on January 1, 1921. Little more than a week later, following Chief James W. Bennett's retirement, McGill was named acting chief. He was officially promoted, becoming the fifth fulltime chief of department, on June 1, 1921. Shown here in 1924, he died, aged 64, on June 30, 1931, of a heart attack reportedly caused by overwork and stress over the ill health of his wife. (MHFM.)

John G. Ford Sr., shown here in 1924 as an assistant chief, joined the department on October 30, 1895, and was made captain in December 1905. He was promoted again in May 1915, becoming the department's first battalion chief. That position was created to provide (by the battalion's being stationed at Engine 3's South Broad Street firehouse) a faster response by a chief officer to fires in South Trenton. His promotion to assistant chief came on September 15, 1918. Named acting chief after Chief Jeremiah McGill's death, Ford officially became the sixth chief of department on August 1, 1931. He retired on May 1, 1935, and died at the age of 80 in 1953. (MHFM.)

Assistant Chief George E. McCrossan, pictured here in 1924, joined the department in 1908. While fighting a house fire on October 16, 1913, he suffered a dislocated shoulder after he fell from a ladder. He was promoted to lieutenant in 1916, then captain in 1918. He was promoted again to battalion chief on January 1, 1921, and later that year made assistant chief. On July 16, 1917, Lieutenant McCrossan was rescued from the third floor of a burning store after he was overcome by smoke. Years later, on February 3, 1927, Assistant Chief McCrossan was the rescuer, saving Russell Schuchardt after dense smoke from burning cardboard overcame the firefighter and he collapsed during a fire at Whitehead Pottery. McCrossan retired in 1935 and died, aged 81, in 1953. (MHFM.)

Roland H. Weigand was also an assistant chief in 1924. Appointed to the department in October 1905, his first promotion—to lieutenant—came in 1911. On January 1, 1917, he was made captain. He was next promoted to battalion chief on January 10, 1921. His promotion to assistant chief came on June 1, 1921. He was named acting chief on May 1, 1935, following Chief John Ford's retirement, and was officially made the seventh chief of department on June 12, 1935. Weigand retired on March 1, 1938, and died, aged 63, in 1940. (MHFM.)

Louis M. Scheidnagel held the rank of battalion chief in 1924. An old Ossenberg Hose volunteer (having started there in 1886), he was among the first paid men hired in 1892. He served at Engine 3 from his first day on the job, through promotions to lieutenant in 1906 and captain in 1907, until he was made battalion chief in 1921. In 1913, Scheidnagel and Capt. John G. Ford Sr. took part in a month-long Fire Department of New York training program, during which time Scheidnagel rode with Truck 24 and Ford with Engine 33 in Manhattan. Scheidnagel was promoted to assistant chief in 1926. The last of the original paid firefighters, he retired in 1928. He died in 1949, aged 86. (MHFM.)

This was the first of the department's searchlight wagons (also known as floodlight trucks or light plants). Capt. Valentine J. Burkhauser is shown admiring the new apparatus after it was put in service in May 1927. Fitted with three 500-watt lights, it was built entirely by department repair shop personnel using a Reo chassis. Shop mechanics made several other rigs around this time, including Trenton's first "emergency car." Carrying specialized equipment like an acetylene torch and battering rams, that vehicle was the predecessor of the civil defense rescue truck that came along decades later and can rightly be thought of as the grandfather of today's Rescue 1. (MHFM.)

Following the successes of their earlier in-house builds, including another searchlight wagon and a fuel delivery truck, shop staff in 1930 took on the challenge of building their own pumper. A Mack chassis, Waukesha motor, and 1,000-gallon-per-minute Hale pump were obtained. They enlarged the chassis, fabricated body elements using recycled American LaFrance parts, and installed locally made hydraulic brakes. The body was painted red, and ladders and other tools were mounted. The completed rig performed so well in underwriters' testing that two other identical engines were built. Shown here putting the finishing touches on one of those in 1931 are, from left to right, shop personnel Raymond C. Ayres, Martin J. Huley, Leo "Lee" P. Pogletki, and Harry L. Brindley. (MHFM.)

The photograph above, taken just after its delivery in 1928, shows the brand-new Mack tractor-drawn 75-foot aerial ladder that was the first apparatus used by Truck 4. Budgeted for following a 1926 request from Chief Jeremiah McGill for a new truck company to better protect the city's growing population and respond to fires in tall buildings, Truck 4 was formed and went in service on August 2, 1928, from the new Fire Headquarters. Truck 1, which had temporarily relocated there when the new building opened on September 12, 1927, then returned to the West Hanover Street firehouse. The new headquarters resulted from an appeal made in 1925 by McGill in which he noted the poor conditions faced by the companies stationed in the old headquarters at 16 Perry Street (such as one toilet shared by 16 men). Several locations, including a city-owned lot at North Clinton and Lincoln Avenues, were considered before 244 Perry Street was finally chosen. Construction of both the new headquarters and the state-of-the-art department repair shop behind it reportedly cost $385,000. The c. 1930 image below shows the crews of Engine 10 and Truck 4. (Above, author's collection; below, MHFM.)

In the 1928 photograph above, Capt. Ervin E. Terry (next to the steering wheel) conducts a training exercise with members of Engine 1. According to the department's 1928 annual report, the apparatus being used is the emergency car that department mechanics built in-house using a Boyd chassis. As department drillmaster, Terry regularly trained firefighters in the drill tower and smoke house at the rear of the new Fire Headquarters. Before these facilities were built, firefighters trained on a 65-foot drill tower erected in 1913 behind the First District Police Station on Chancery Lane. Standard drills included practice using pompier scaling ladders, life nets, and "smoke masks" (crude breathing apparatus). Instruction was also given on basic first aid techniques of the time, such as the Schafer prone-pressure method of resuscitation (in which a victim was laid face-down and compressions made to the middle of his back). The 1929 photograph below shows another crew of firefighters in the drill yard behind headquarters. (Both, author's collection.)

Cleveland No. 69 (Opposite)-February 23, 1930

10:40 A.M., returning at 10:50 A.M., still.
Notified to go get cat out of tree on Cleveland
Ave. near Walnut Ave. Cat rescued by Dura,
Stackhouse and Gaertner assisted by the Company
Used one extension ladder. Walter Stackhouse
was bitten on the right hand by the cat. Noth-
ing used by Department. No fire damage. Eng.
Co. #6 answered alone. Officer in charge,
Lt. Ecker.

For decades in the pre-digital age, condensed details about every emergency response made by the department were typed onto three-by-five-inch index cards and filed alphabetically by location. Multiple cards fastened together were used to document larger fires and major incidents. This "database" eventually grew to thousands of cards about all kinds of calls, from tragic events like fatal fires and drownings to more humorous incidents like this February 1930 cat rescue. (MHFM.)

Fire Headquarters,
Trenton Fire Department,
Trenton, N.J.

May 20, 1931.

To Commanding Officers of Companies:

Below please find extract from two letters of commendation, one from the Radio-Keith-Orpheum and the other from the Trenton Trust Company, both dated May 6, 1931, in connection with splendid work accomplished in fighting fire at the Capitol Theatre, May 2, 1931:

Radio-Keith-Orpheum Corporation
R.K.O. Capitol Theatre,
Trenton, N.J.

"I desire to express through you my apprecia-tion for the efficient work of both the Fire & Police Departments at the Capitol Theatre fire. "You have just reason to take pride in the high standard and devotion to duty displayed by both superior officers and men in the ranks. "Your deputy, Mr. Harry Kemler, was one of the first to arrive at the scene and he also gave every possible aid to help reduce the property loss."

Yours very truly,
/s/ J.M.Brennan.

- - - - - - - - - -

Trenton Trust Company,
28 West State St.,
Trenton, N.J.

"Through you, I want to express my apprecia-tion of the splendid job the Fire Department did in controlling the bad fire at the Capitol Theatre. As you know, we are the owners of the Hotel Windsor property and, naturally, interested. From all that I can learn, the Department did remarkable work in keeping it confined to the one building. I thought that I would like to write you, commending it."

Very truly yours,
/s/ H.A.Smith,
President.

Respectfully,

Acting Chief Fire Dept.

Assistant Chief John Ford, serving as acting chief while Chief Jeremiah McGill was out of town, issued this memo on May 20, 1931, praising firefighters for their work battling the four-alarm blaze that destroyed the RKO Capitol Theatre on May 2. Located at 18 South Broad Street, the building opened in 1867 as the Taylor Opera House. Faulty wiring sparked the fire. Manager James Brennan was lauded for the way he calmly evacuated over 1,000 patrons from the theater as staff dropped an asbestos curtain and opened skylights in an effort to contain the fire backstage and draw flames to the roof, away from the fleeing crowd. The first alarm was sounded from Box 8112 at 11:13 p.m. Just 11 minutes later, the fourth alarm was ordered. Several neighboring businesses were threatened, but firefighters ultimately prevented the flames from spreading. (MHFM.)

Members of Engine 1 and Truck 1 are shown here in their quarters in September 1930. The 1913 Webb truck on the right was refurbished, with a new Mack tractor installed, in 1926. (MHFM.)

Engine 6 members, having relocated in 1927 to Truck 3's old firehouse on South Olden Avenue, are shown here around 1933 with their American LaFrance pumper. From left to right are Lt. David J. Burke, William Megules, John Seitz, William Buckley, George Woolverton, and Earl Campbell. (MHFM.)

This three-alarm fire destroyed several wood-frame storehouses and badly damaged the main plant, a three-story brick structure, at Trenton Brass and Machine Company on Prospect Street on March 24, 1933. The first alarm (Box 3233) was struck at 1:00 p.m., followed by the second at 1:13 p.m. and the third at 1:22 p.m. Firefighters George Anderson and Peter Schafer, both of Engine 5, suffered burns to their hands and faces when they were overrun by flames after fire burned through their hoseline. Several houses on nearby Dale Street were damaged by the intense radiant heat. (MHFM.)

Chief Roland Weigand is shown on March 19, 1936, checking conditions after heavy rains, combined with warm temperatures that melted ice and snow, caused the Delaware River to overflow its banks and flood the Glen Afton neighborhood. Firefighter George Woolverton mans the oars. (MHFM.)

Two weeks of heavy rains in January 1937 resulted in widespread flooding along the Ohio River. In Louisville, Kentucky, 15 inches fell in 12 days, causing the river to crest 30 feet above flood stage and flood over 60 percent of the city. Trenton city manager Paul Morton, a Louisville native, returned home on January 28 in answer to pleas for help from Louisville's mayor. Tasked with helping manage public safety operations, Morton called back to Trenton for backup. Nine Trenton firefighters and fire alarm operator Raymond Apgar volunteered to aid Louisville's exhausted firefighters. Shown on January 30 preparing to board an American Airlines flight from Newark Airport are, from left to right, Edwin Moore, Jacob Hiel, Fred "Fritz" Miller, Apgar, George McClellan, Deputy Chief Thomas Phelan, Capt. Thomas Dovgala Sr., Lt. William Miller, Emerson Hill, and Anthony Pyontek. When they reached Louisville, the Trentonians entered the flooded city via a pontoon footbridge made of empty whiskey barrels. While Apgar worked to maintain the local fire alarm system, the others were assigned to companies alongside Louisville crews and other mutual aid firefighters from Pittsburgh and Indianapolis. Moore and Hiel were hailed as heroes after they saved a woman and child from a burning building on February 5. Ten people died in that three-alarm blaze started by a gas explosion. Miller (treated for pneumonia upon his return to Trenton) and Hill were also praised for carrying an elderly heart attack victim down six flights of stairs to get medical aid. The weary group, returning by train, was welcomed home in Trenton on February 9. (MHFM.)

At 8:11 p.m. on July 11, 1937, during a violent thunderstorm, four engine companies, Trucks 1 and 2, Searchlight 2, the department's emergency car, and Battalion Chief Valentine J. Burkhauser were dispatched (Box 8461) for an ADT fire alarm at People's Brewing Company at Lalor and Lamberton Streets. It would turn out to be a false activation caused by a lightning strike. But as Engine 1 was on Centre Street en route to the assignment, it collided with Engine 7, approaching Centre Street from Cliff Street. Firefighters riding the rear steps were thrown in all directions as the apparatus came together. Engine 1 then hit an iron fence, while Engine 7 continued another 100 feet before acting lieutenant William Hartz and driver Robert Van Hise managed to bring it to a halt. A total of nine men were hurt. The worst, with chest and shoulder injuries, was John Szagany. He was hospitalized, together with Richard Behrmann, Peter Downs, Peter Howard, John Seaman, and Van Hise, who suffered various fractures, cuts, and bruises. Lt. Adolph Doldy and Firefighters George Weeks and John Powers were sent home after they received medical treatment. All nine men would eventually recover and return to work. Both apparatus—Engine 1's department-built Hale and Engine 7's American LaFrance—were badly damaged but were put back in service after over $5,000 in repairs. Shown above with the wreckage of Engine 1 is Battalion Chief Burkhauser. Tragically, one of the survivors of this crash, Weeks, was killed in another apparatus accident in 1951. (MHFM.)

This three-alarm fire on January 25, 1938, gutted nine businesses at 99–115 South Broad Street and 35–39 East Front Street, where Trenton's old Washington Market once stood. Wooden partitions between stores did little to slow the blaze, and the common basement was a mass of flames when first-due companies arrived after the first alarm from Box 8112 was sounded at 6:56 p.m. At least 12 firefighters received medical treatment after being overcome by the dense smoke. Department records show that nearly three dozen 2½-inch hoselines were put in service. (MHFM.)

This photograph, believed to have been taken on September 28, 1938, in Stacy Park behind the state house, shows Truck 1's new apparatus being tested prior to the city accepting delivery from the manufacturer, Peter Pirsch of Kenosha, Wisconsin. Equipped with a 100-foot aerial ladder, the rig was put in service on September 30, 1938. At $17,648, Pirsch's price was lower than bids submitted by American LaFrance and Seagrave. Trenton bought several more Pirsch ladder trucks over the years, including one with an 85-foot aerial in 1948 that replaced Truck 4's 1928 Mack. The Mack was then sold and used into the mid-1950s by Harrisonburg Fire Department in Virginia. (Author's collection.)

```
7 Pm.   Circuit test, O.K.

          Box # 6316 at 7.57 Pm.
          Engine #4.- 8.08 Pm.
          Engine #6.- 8.30 Pm.  & taps                    Outer house rear of
          Truck # 3.- 8.13 Pm.                            11 Parker Ave.
          Lite # 2.- 8.28 Pm.

7.57 Pm. Also calls on # 11 E.T. from State & Logan  for same fire.

8.13 Pm.  Calls pouring in from all sections of the East, regarding a supposed
          meteor falling around Groveville and vicinity. Persons excited were
          asking if it was safe to remain in Trenton, and how much danger was
          there.  We received calls from the Associated Press in New York, the
          United Press. and Radio stations in New York and Philadelphia.  N.Y.
          and Delaware State Police also making inquiries, Public Service and
          Telephone Company reports being swamped also with calls for information.
          Upon checking these rumors we learned that the source of the scare
          came from Radio station WABC, from a dramatization of the play entitled
          The War of the World, and that during the discourse the author had ment-
          ioned that Trenton, and Vicinity was being destroyed.  We informed
          persons calling that they need not be alarmed, and helped quiet the
          residents of Trenton and vicinity.
10.23 Pm.
          Manager Morton, calling, and asked us to continue to pacify persons
          calling, also Manager called New York Radio Station WABC. asking
          them to deny that any catastrophe had happened around Trenton or
          vicinity.  TRUNKS STILL BUSY.

11 pm.   Circuit Tests, O.K.
```

In the days before every home had a television, families gathered around the radio to be entertained by variety shows, scripted comedies, and full-cast dramas. On October 30, 1938, Orson Welles and his Mercury Theatre actors staged their infamous audio version of H.G. Wells's *The War of the Worlds*, using a series of realistic-sounding news bulletins to tell the story of a Martian invasion occurring just a dozen miles north of downtown Trenton. Listeners who tuned in late missed announcements at the start of the 8:00 p.m. broadcast identifying it as fiction. So real did these reports sound that when the aliens launched their murderous onslaught at Grovers Mill with death rays and poisonous gas, many listeners across the country—particularly those closest to the imaginary battle—thought it real. Sylvester "Bud" Zimmerman and Leon Smith were the operators (forerunners of today's 911 dispatchers) on duty that night in Trenton's police and fire communications bureau on West Hanover Street. What had been a quiet night, with only one small fire reported thus far, suddenly turned into their busiest night ever. As Zimmerman recalled in an interview years later, the switchboard "lit up like a Christmas tree" and the pair were inundated with telephone calls from terrified residents asking if it was safe to stay in the city. Other calls came from New York and Delaware state police barracks, the Associated Press, and newspapers as far away as London all wanting information on the invasion. At first unclear what was happening but then learning of the hoax (Orson Welles himself, at the conclusion of the broadcast, described it as a radio version of a Halloween prank), the communications bureau staff spent several hours trying to calm residents. Shown here are entries about that night made by Zimmerman and Smith in the bureau's fire alarm log. Thousands of typed fire alarm log pages, documenting the daily activities of the department from 1930 through 1982, are preserved at the Trenton Free Public Library. (TFPL.)

At 3:50 p.m. on March 30, 1940, Engines 9 and 4, Truck 3, and Battalion Chief George A. Weigand were dispatched on Box 5426 to the Pennsylvania Railroad yard on Mulberry Street for a fire in a boxcar reportedly started by children playing with matches. Five minutes later, Box 4227 sounded, and Engines 5 and 10 and Truck 4 were sent to a different job. Firefighters under the command of Deputy Chief Valentine J. Burkhauser are shown here working to put out that second blaze at the Robinson Process Company's coal nugget manufacturing plant on Brunswick Avenue, near Vine Street. (MHFM, photograph by J.V. O'Neill.)

Shown here is the rebuilt apparatus put back in service as Truck 2 in July 1941. The rig—originally constructed for Truck 2 in 1917 by Couple-Gear, then refurbished in 1925 with a new American LaFrance tractor installed—was taken out of service on September 22, 1940, by Chief Thomas Phelan. For 10 months, department repair shop superintendent Frank Webb and his crew (Raymond Ayres, Harry Brindley, Martin Huley, Lee Pogletki, and Sturley Garrison) reconstructed the rig, spending just $3,500. Shop painters John Entwistle and Louis Black then gave it a brand-new finish. While the apparatus was out of service, Truck 2 members responded to fires in a reserve engine. (MHFM.)

The woodworking plants and lumber yards of the A.K. Leuckel and Wilson and Stokes companies in the 600 block of South Broad Street were ravaged by this four-alarm blaze on August 29, 1941. The first alarm (Box 8244) was transmitted at 4:20 p.m. The fourth alarm was struck at 4:54 p.m. A group of fire buffs and off-duty fire dispatchers—William Brennan, Edwin Fisher, Leroy Fisher, Ernest Hubscher, and Walter Parker—was among the large crowd of spectators on scene. They pooled their money to buy coffee and sandwiches for the weary firefighters that hot, humid evening and went on to organize the Signal 22 canteen service, which still responds to fires today. (MHFM.)

Truck 4 firefighter Daniel Tomasulo suffered minor burns while operating at this two-alarm blaze at the Park Theater on July 6, 1947. Another firefighter, John Breza of Truck 2, stepped on a nail. Responding to reports of an explosion at 10:58 a.m., first-due Engine 7 arrived to find heavy smoke showing from the movie theater at Anderson and Washington Streets. Box 7373's second alarm was struck at 11:02 a.m. A close look reveals Rex, the proud canine mascot of Engine 7, sitting atop the apparatus as his crew stretches a hoseline. (MHFM.)

One firefighter was treated for smoke inhalation after companies led by acting deputy chief John T. Dempster Sr. extinguished this blaze at 685 South Broad Street on August 15, 1947. Just three days earlier, Dempster and three other firefighters had to be revived after they were overcome by smoke during a two-alarm basement fire at the S.P. Dunham department store at 11 North Broad Street. Dempster, the department's drillmaster from September 19, 1947, until his June 1, 1959, retirement, also served as Mercer County's fire marshal for nearly 14 years. The county's fire academy, which opened in 1986, is named in his honor. (MHFM.)

Shown here around 1947 are members of Engine 1 and Truck 1. From left to right are John Bayen, Alvin Slack, Edward Michaloski Jr., George Weeks, Capt. Adolph Doldy, Michael Mune, William McGuire, Edward Hicks, Michael Cimbala, and Alfred Walton. The department's first apparatus with a fully enclosed cab, Engine 1's 1941 Buffalo pumper, is also shown along with Truck 1's 1938 Pirsch. Department records show the Buffalo's first run was to a small fire, reported at 2:29 p.m. on June 3, 1941, in a meat packing business at Bloomsbury and Fall Streets. (MHFM.)

During this three-alarm blaze on December 1, 1947, firefighters waded through the icy Assunpink Creek to bring ladders and 2½-inch hoselines into play at the rear of the four-story building at 126–128 South Broad Street shared by the New Jersey Floor Covering and D. Wolff Furniture stores. The fire, which started in the basement of the furniture store, was reported from Box 8114 at 7:12 a.m. The second alarm was transmitted at 7:34 a.m., followed by the third alarm at 7:48 a.m. Several neighboring businesses sustained smoke damage but were otherwise saved as firefighters managed to hold flames to the basement and first floor of the original building. Apparatus operating at the scene included Engines 1, 2, 10, 3, 5, 6, 7, and 4, Light Plant 1, and Trucks 1, 4, and 3. (MHFM.)

Just two months after the three-alarm fire shown on the opposite page, the building housing New Jersey Floor Covering and D. Wolff Furniture was fully destroyed by this general-alarm inferno on February 10, 1948. The above image (and the one on page 2) shows the massive efforts undertaken to fight the flames, while the photograph below shows the ice-encrusted ruins of the stores and the neighboring businesses that were damaged. An oil stove used by contractors repairing damage from the previous fire was blamed for starting this blaze. The first alarm was sounded at 4:53 p.m.; by 5:13 p.m., the fifth alarm had been struck. With every city company already working, volunteer firefighters from Hamilton, Ewing, Morrisville, and Yardley were called to the scene to help. The body of the contractors' watchman (believed to have been killed while trying to fight the flames) was later recovered from beneath debris that had fallen into Assunpink Creek. (Both, MHFM.)

It was common in this era for large crowds of spectators, sometimes numbering in the thousands, to gather at fires in Trenton, particularly at bigger daytime blazes in the downtown business district. This three-alarm fire on February 14, 1948, gutted the top two floors of Binder's department store at 132 East State Street. It was first reported from Box 6112 at 12:18 p.m. Recognizing the severity of the fire, Deputy Chief Thomas Dovgala Sr. ordered the second and third alarms at 12:28 p.m. To access the building's west side, firefighters stretched hoselines into the neighboring graveyard of First Presbyterian Church. Shortly after 5:00 a.m. the next day, just hours after firefighters had left, a rumbling was heard in the ruins. As Truck 4 was temporarily short a man, Truck 1 was sent to investigate. What happened next was recalled years later by Chief George Weigand: "Just as Truck 1's captain and four men were about to enter, the interior collapsed with a roar. Had they arrived but seconds earlier—which surely would have been the case if there hadn't been that slight delay due to the switching of companies—those firemen would have been crushed to death." (MHFM.)

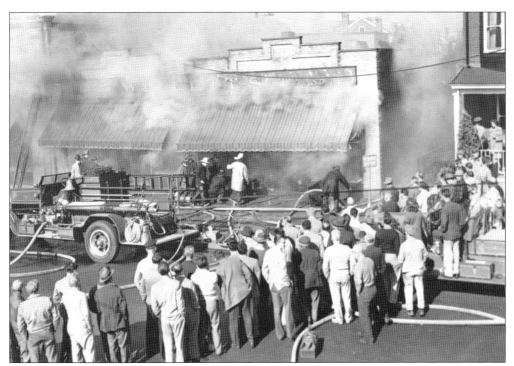

It was 8:52 a.m. on October 4, 1948, when Gamewell fire alarm Box 4327, at the corner of Brunswick and North Olden Avenues, was pulled for this blaze at Willett's discount store at 901–903 Brunswick Avenue. The fire, having started in the cellar, had burned through flooring in the center of the store before firefighters arrived. As heavy smoke churned from the structure, a second alarm was ordered at 9:00 a.m. At one point, Capt. John T. Dempster Sr. fell through the hole burned in the floor. Landing in three feet of water pooled up in the cellar, he was not injured. (MHFM.)

All four truck companies were put to work at this three-alarm blaze on December 11, 1948, at the Utility Service Supply Company's household appliance store at 142 South Broad Street. Firefighters were first alerted by an alarm sounded at 9:16 p.m. from Box 8123, several blocks away at South Warren and Lafayette Streets. The third alarm was transmitted at 9:26 p.m., after first-arriving companies found that flames had already spread from the basement to upper floors via an elevator shaft. Neighboring buildings housing the Workingman store and the Original New York Furrier Company were damaged before the fire was brought under control. (MHFM.)

Four brand-new Ward LaFrance pumpers are shown on January 21, 1949, undergoing insurance underwriters' testing at the Trenton Marine Terminal (which was built in 1931 along the Delaware River off Lamberton Street, south of Riverview Cemetery). The rigs, each equipped with a 1,000-gallon-per-minute pump, were put in service a few weeks later as the new Engines 2, 3, 4, and 10. They cost the city a combined $52,000 and the trade-in of two old 1923 American LaFrances. Five more Ward pumpers (including four delivered in December 1953) were put in service as Engines 1, 5, 6, 8, and 9 in 1954. Engine 7 finally got its own new Ward in 1957. (MHFM.)

Capt. Conrad Lenius was on duty at Truck 3 on May 8, 1950, when the Gamewell fire alarm register punched out numbers for the corner of Willow and Bank Streets (Box 2131). Truck 3 did not ride on that box, so it was not until later that Lenius learned of this fire that had engulfed the first floor of his North Willow Street home, trapping his wife and lodgers. Lenius's wife was rescued from a second-floor window. Two tenants were also saved, but one died of her burns the next day. Another lodger was killed when she jumped from the third floor before firefighters arrived. (MHFM.)

Four

1951–1990

On May 11, 1949, Gov. Alfred Driscoll signed legislation allowing municipalities to give firefighters a 56-hour work week through creation of a three-platoon system, replacing the 84-hour work week under the old two-platoon system. But Trenton did not introduce the three-platoon system until January 1951, despite the fact that Trenton firefighter Daniel Tomasulo, as state president of the New Jersey Firemen's Mutual Benevolent Association union, drafted the bill on which the new law was based. Fourteen years later, Trenton began phasing in the four-platoon system and its 42-hour work week. But that change (which was not fully implemented until 1970) came at a price—Engine 4 was put out of service on January 12, 1965, and officially disbanded the next day, its members reassigned to other companies to start the transition without a need for any new hires. Truck 3, in turn, was reassigned from Brunswick Avenue to take Engine 4's place on North Clinton Avenue.

In 1954, the department finally got its own radio system, and two-way radios were installed in all apparatus. Interestingly, the department first experimented with similar technology way back in 1921, when "wireless telephoning" equipment was installed at headquarters and in Chief Jeremiah McGill's car. Tests conducted throughout Trenton proved that instant, clear communication was possible. New York–based periodical *Fire Service* in October 1921 praised McGill for taking "a great step forward in scientific firefighting," noting how radio could save time calling extra resources to large fires and be used to relay vital information when severe weather disrupted regular telephone service. But the equipment was removed just a month later after Trenton newspapers criticized the radios' cost and the proposed hiring of more staff to operate the system at headquarters.

In December 1954, the assistant chief rank was resurrected to designate a second-in-command for the department and clear successor to the chief. On January 1, 1955, Deputy Chief George A. Weigand was sworn in as the new assistant chief. The department's rank structure was modified again in 1964, when the second-in-command position was changed from assistant chief to deputy chief, and the old deputy chief positions became battalion chiefs.

Another major change came at the start of 1959, when George Weigand, by then having been promoted to chief, renamed all truck companies to ladder companies. Weigand explained at the time that he believed the name switch necessary to ensure clarity when transmitting orders by radio.

Chief Thomas J. Phelan (center) is shown in July 1951 inspecting the new fuel truck built in-house on a Chevrolet chassis by shop staff. The truck, housed at Engine 7's quarters, carried 300 gallons of gasoline. With Phelan are mechanic Martin Huley (left) and Capt. Frank Webb, chief mechanic. It was because of his prior experience as an automotive mechanic that Webb was hired in 1909 to drive the new 4-cylinder Webb engine (their shared name being just a coincidence). (MHFM.)

Shown here in August 1951 is the committee put together to select the winners of the department's yearly fire prevention essay contest. From left to right are (seated) Capt. Frank Webb and Chief Thomas J. Phelan; (standing) Captains Anthony Pyontek, John Tresansky, and Allen Shipley. The contest, open to graduating high school seniors, was held annually from 1946 through 1960. Each winner received a $300 college scholarship check from the firefighters' union. (MHFM.)

Arriving in response to an alarm turned in from Box 8132 at 7:45 a.m. on January 7, 1953, first-due firefighters found heavy smoke pumping from 51–53 Cooper Street. Seven adults and 12 children had already fled the homes, including a woman who had jumped from a second-floor window after dropping her infant daughter to neighbors below. Water supply problems caused by poor pressure from the two closest hydrants resulted in the second and third alarms being called. (MHFM.)

Pictured here in front of the Trenton War Memorial in December 1953, just after their arrival in the city, are two brand-new Pirsch-built ladder trucks. The apparatus, each with an 85-foot aerial, were assigned to Truck 2 and Truck 3 and placed in service in January 1954. (MHFM.)

Flames swept through the Cook Ceramic Manufacturing Company at 500 Prospect Street on May 27, 1954. The fire started in a second-floor storeroom and rapidly spread, sending the factory's 175 employees fleeing. Having been dispatched at 10:57 a.m. to investigate an activated ADT fire alarm, first-due companies arrived to find heavy fire showing. Box 3233's second and third alarms were called together at 11:02 a.m. Crews from seven engines and two trucks operated on scene. (MHFM.)

Twelve firefighters were injured battling this general-alarm blaze on January 6, 1956. The fire started in the cellar of the Morlee clothing store and rapidly spread to all three floors above before burning through the roof of the business at 20 East State Street. The first alarm (Box 2112) was called at 7:32 p.m. At 8:02 p.m., the fifth alarm was ordered. Neighboring businesses, including Dunham's department store, sustained smoke and water damage. Not long after being released from East State Street, several companies responded to a South Broad Street house fire and rescued a woman trapped on the second floor. Four firefighters were hurt during that blaze. (MHFM.)

General-alarm fires reported just hours apart on March 14, 1956, destroyed the original St. Mary's Cathedral (above) and badly damaged Jefferson Elementary School (right). Having stood at North Warren and Banks Streets since its 1871 dedication, the cathedral was deliberately set ablaze. The arsonist (who was later arrested and confessed) ignited a fire in the chancery office. Flames, fueled by gasoline, rapidly spread. Monsignor Richard Crean became trapped while trying to warn others and died together with two housekeepers. Several priests were rescued via aerial ladders. A life net was used to save another who jumped from a third-floor ledge. Box 2123's first alarm was sounded at 4:31 a.m. After the fifth alarm was struck at 4:50 a.m., even more help was called in from nearby volunteer companies. St. Mary's was still smoldering when smoke was spotted coming from a third-floor closet at the Brunswick Avenue school, and Box 4224 was transmitted at 9:05 a.m. Suburban volunteers still covering city firehouses were among the first firefighters to arrive. More alarms were quickly sounded as over 700 students and teachers were evacuated to safety. (Both, MHFM.)

The St. Mary's arsonist (later diagnosed as mentally ill) struck again on December 16, 1956, when he started gasoline-fueled fires in four more churches. Two of these burned themselves out, causing minor damage, and another was quickly controlled by firefighters. But the fourth blaze, seen here, badly damaged the First Methodist Church at 15 South Broad Street and destroyed the church's two-story school. Five firefighters narrowly avoided being crushed when the school roof collapsed. Reported at 4:29 a.m. (Box 8112), this fire went to a general alarm. Thanks to evidence found at the scenes, police identified and arrested the arsonist the next day. A copycat arsonist (also mentally ill) set fires at two Trenton churches and a synagogue just days later and was also arrested. (MHFM.)

Capt. John Toth (center) and Firefighters Joseph Krasnovsky (left) and Joseph Reilly are shown in March 1957 reviewing the different breathing apparatus in use by the department at the time. Reilly is wearing an all-service gas mask, manufactured by MSA. (Author's collection.)

This general-alarm blaze ravaged several buildings at the corner of North Broad and East Hanover Streets in the heart of Trenton's business district on October 27, 1957. The first alarm (Box 3112) was transmitted at 9:12 p.m., followed by the second at 9:22 p.m., the third at 9:31 p.m., the fourth at 9:57 p.m., and the fifth at 10:13 p.m. Special-called to the scene at 10:23 p.m. was additional help from the Capitol View, Lawrence Road, and Prospect Heights volunteer fire companies. The blaze started in the basement of Hamilton Jewelers at 24 North Broad Street and rapidly spread to Norman's Gifts next door. A half-dozen other businesses sustained damage before flames were brought under control. The heat in Hamilton Jewelers' basement was so intense that Deputy Chief Francis "Frank" Apgar Sr., who arrived with first-due crews, later told reporters, "It was so hot you could hear the hair singeing around your ears." (Above, author's collection; right, MHFM.)

The above image shows Assistant Chief George A. Weigand (left) and Chief Thomas J. Phelan sometime after Weigand's promotion to assistant chief on January 1, 1955, and before Phelan's death on January 12, 1958. Phelan joined the department in 1914. He was promoted to lieutenant in 1921, then captain in 1923, and battalion chief in 1928. He became assistant chief in 1935. A year later, his title changed to deputy chief due to a rank restructuring. He was sworn in as the department's eighth chief on March 1, 1938. After Phelan's fatal heart attack at age 65 (having been allowed to stay working past his 65th birthday), Weigand served as acting chief until he was officially made the ninth chief of department on January 27, 1958. Below, Weigand receives a new helmet from Capt. Robert Kerr in April 1958. Having also come on the job in 1914, Weigand was promoted to lieutenant in 1923, then captain in 1928. He was elevated to battalion chief in 1938, then deputy chief in 1943. Retiring at 70 in early 1963, he died aged 92 in 1985. (Both, MHFM.)

Just hours after a Mercer County grand jury indicted the St. Mary's firebug on arson and murder charges for the cathedral's 1956 burning, city firefighters found themselves battling this three-alarm blaze in the Workingman store at 144 South Broad Street on November 26, 1957. Box 8114's first alarm was sounded at 11:12 p.m. The second alarm was struck two minutes later, as responding companies spotted heavy smoke in the air. The third alarm was transmitted at 11:19 p.m. Almost four decades earlier, on March 15, 1919, another three-alarm fire gutted the same building, which back then housed the Smith Brothers furniture store. (MHFM.)

Hampered by sub-freezing temperatures, firefighters had a difficult time tackling this general-alarm blaze on December 20, 1958. Not long after the Style Shop had been locked up for the night, an explosion blew out plate-glass windows of the business at 136–138 South Broad Street. Box 8114 was transmitted at 9:09 p.m. as thick smoke rolled out. By 9:18 p.m. every on-duty city firefighter had been dispatched. Suburban volunteer companies were then called in to help. Four firefighters were injured. Neighboring properties affected by previous fires (like the one above) once again sustained smoke and water damage. (TFPL.)

Firefighters worked from adjacent roofs during this general-alarm fire that gutted Mercer Bag and Burlap Company at 311 Bloomsbury Street on March 8, 1959. Police checking a smoky haze in the area reported the blaze via radio just as someone turned in Box 8234 at South Warren and Bridge Streets at 1:29 a.m. (This neighborhood no longer exists as it was then, having long since been redeveloped for the Trenton Freeway, state government buildings, and parking lots.) Firefighters later estimated that they flowed half a million gallons of water to extinguish the blaze. (MHFM.)

In January 1957, Trenton's Civil Defense Council took delivery of this rescue truck. Responsibility for manning the apparatus later shifted from civil defense volunteers to a team of firefighters sent to special training like the New Jersey State Police's heavy rescue school in Hammonton. Shown in May 1959 with the rig are, from left to right, Francis Apgar Jr., Peter SaFranko, Chief George Weigand, George McQuillan, Capt. Joseph Shelby (drillmaster), Edward Gore Jr., public safety director William Waldron, William Noon, and John Oakley Sr. Although this cross-manned truck was called "Rescue 1," the actual company designated Rescue 1 was not organized until 1988. (MHFM.)

Chief George Weigand and chief mechanic Martin Huley are shown in 1959 with five rigs built by department mechanics: two 1931 Hale pumpers, searchlight wagons assembled in 1941 and 1945, and the fuel truck put in service in 1951. The photograph was taken just prior to the Hales being traded in as part of the deal for new 1,000-gallon-per-minute Maxim pumpers for Engines 2 and 10. The 1930 shop-built Hale was previously traded in under the terms of the 1957 purchase of Engine 7's new Ward LaFrance pumper. Two of the Hales, bought at auction by the State of New Jersey, saw several more years of work at Marlboro Psychiatric Hospital and Leesburg State Prison. (MHFM.)

This March 13, 1960, three-alarm fire gutted the vacant building at 541 East State Street previously used as a studio for radio station WTTM. Firefighters answering Box 6117's first alarm at 7:25 p.m. arrived to find fire showing from the roof and thick smoke, generated by burning acoustic panels, pouring out. The second alarm was struck at 7:43 p.m., followed by the third at 7:52 p.m. (TFPL.)

As the Cold War with the Soviet Union heated up, Trenton firefighters prepared for the worst. In April 1959, Chief George Weigand and several other officers attended training on chemical warfare. In January 1960, more firefighters learned how to handle radioactive materials emergencies. And in April 1961, civil defense funds equipped each engine company with its own Geiger counter. Shown here are firefighters taking part in a Cold War training exercise (the scenario supposedly involving a burning truck carrying an "atomic weapon") on November 19, 1960, on Kossuth Street. (MHFM.)

Memories of the 1956 general-alarm fire at Jefferson Elementary School were immediately brought to mind when Trentonians learned that 92 children and three teachers had died in a fire at Our Lady of the Angels School in Chicago on December 1, 1958. Because of that tragedy, Trenton committed to installing fire alarm stations in every school. It was not until April 1961 that alarms were finally in service at all 40 public and parochial schools citywide. Shown here at Blessed Sacrament School on November 21, 1960, testing the first alarm are, from left to right, police chief Louis F. Neese, Rev. J. Arthur Hayes, Chief George Weigand, and public safety director William Waldron. (MHFM.)

Lots of families have strong departmental ties. Many firefighters followed in their fathers' footsteps, some lucky enough to serve alongside their dads. Brothers have worked together. Chief George Weigand was a nephew of Chief Roland Weigand. Fire museum cofounder Edward Gore Jr. is shown overhauling a two-alarm fire in a jewelry store on December 14, 1960. His father, Edward Sr., and grandfather George were both city firefighters. Before Edward Jr. retired as a captain in 1990, he worked with his sons Jeff and Greg. Deputy Chief Francis Apgar Sr., before retiring in 1965, fought fires with sons Francis Jr., Richard, and Walter. Eight years after Firefighter Manuel Rivera's 2009 line-of-duty death, his daughter Natasha joined the department. (MHFM.)

Linwood Collins was Trenton's first African American firefighter. A Marine veteran of World War II who saw combat in the Pacific, he joined the department in January 1957 and was assigned to Engine 1. In March 1961, Collins searched through blinding smoke to find and rescue a two-year-old girl from the second floor of a Prospect Street home. Ill health forced Collins to give up firefighting in 1962. He died in 1992 aged 66. Other significant department firsts came in 1977, when Homer Rawson became the first African American firefighter promoted to captain; in 1982, when Gerry Castro was appointed as the first Hispanic firefighter; and in 1995, when Alisa McNeese was hired as the first female firefighter. (MHFM.)

Ladder pipes are shown flowing water on this three-alarm fire at the Pennsylvania Railroad's station at South Clinton and Fairview Avenues on April 16, 1962. Replacing the old Clinton Street train depot, the building opened to passengers in December 1891. The fire, started by an electrical fault, gutted a basement communications room before spreading through ducting to the attic. Firefighters were alerted at 2:29 a.m. after police, checking an activated burglar alarm, discovered the blaze. (TFPL.)

Shown after his 1958 promotion to the second-in-command assistant chief's post is Thomas J. Dovgala Sr. A veteran of World War I, he joined the department in 1921. He was made lieutenant in 1928, then captain in 1931, and then deputy chief in 1947. Following Chief George A. Weigand's retirement, Dovgala was sworn in as the department's 10th chief on February 22, 1963. Dovgala retired on April 10, 1965. Because next-in-charge Deputy Chief Francis Apgar Sr. had his own retirement planned for a few months later, he declined the chief's job. Battalion Chief Christopher Reilly was named acting chief to serve until civil service testing could be held to determine the next chief. (MHFM.)

Seven alarms of manpower and apparatus were needed to combat this inferno in the four-story, block-long building at 205 North Willow Street on March 30, 1965. Once used as a mattress factory by Bloom and Godley, most of the structure was in use as a storehouse for the Hy-Grade Pretzel and Food Company. The section fronting on Fowler Street was used by the state labor department to store documents. Box 2134's first alarm was transmitted at 7:37 p.m. Battalion Chief Daniel P. George called the second alarm at 7:43 p.m. Third and fourth alarms, bringing every on-duty city firefighter to the scene, were struck at 7:46 and 7:56 p.m. Before the fire was declared under control at 9:27 p.m., three more alarms of suburban volunteer companies and reserve apparatus manned by recalled off-duty city firefighters were put to work. With every nearby hydrant in use, additional water was drafted by three engines from the Delaware and Raritan Canal Feeder. Chief Thomas Dovgala, who had been using accumulated leave prior to his April 10 retirement, returned to work to help acting chief Christopher Reilly. Thirteen firefighters were injured, including several who braved the intense heat to lower the aerials and move Ladders 1 and 4 to safety before the building's walls collapsed and they were buried. Ladder 1's 1938 Pirsch—paint blistered, ladder charred, and lights melted—remained out of service several days for repairs. Its replacement, a new $53,000 Pirsch delivered in May 1966, was fast-tracked by the city. (MHFM, photograph by Warren Kruse.)

Firefighter John Babice was hospitalized for treatment of smoke inhalation after part of the ceiling fell on him and others as they fought this general-alarm fire on December 22, 1965, at Hiel's Bowling on Whittaker Avenue. Reported at 12:30 p.m. (Box 7334), the fire escalated to four alarms in just 20 minutes. The six-lane bowling alley had been closed 18 months and was up for sale. The building had once been operated as Padderatz Hall, where dances like the old Lincoln Hook and Ladder Company's annual sociable was held in November 1887. (MHFM, photograph by John Pietras.)

Having served 11 months as acting chief, Christopher Reilly returned to his battalion chief duties after Raymond A. Schwab Sr. was sworn in as the department's 11th chief on February 7, 1966. Schwab, seen here, had wanted to be a police officer, but memories of his 12-year-old sister Jennette convinced him to become a firefighter in 1938. It was August 2, 1929, when two dozen bystanders were burned by a fireball of gasoline that ignited after a truck overturned at Centre and Cass Streets. Jennette Schwab and another child died of their injuries. Promoted to captain in 1953, Schwab in 1961 was made deputy chief (later retitled battalion chief). He died, aged 61, in 1971, six weeks after suffering a heart attack while fighting a three-alarm fire. (MHFM.)

Civilian public relations officer Meredith Havens is seen here (white helmet) during a general-alarm fire at Crane Company's ceramics warehouse on January 25, 1967. Firefighters sent to the building at 309 North Clinton Avenue at 6:42 p.m. to check an activated fire alarm found a working fire and quickly called for help. After Box 1523's fourth alarm was ordered at 7:43 p.m., three more engines staffed by off-duty city firefighters and suburban volunteers were requested. One of those volunteers, Prospect Heights' Joseph Lenarski, narrowly avoided being killed by outrunning a large section of wall as it fell. The same site—by then vacant, Crane having ended its Trenton operations in 1970—was destroyed by another three-alarm blaze on May 26, 1971. (MHFM.)

Two alarms were transmitted for this blaze, which destroyed an old warehouse on Parker Avenue on February 1, 1968. First-due companies were alerted at 1:58 a.m. after patrolling police officers spotted flames coming from the three-story warehouse and a one-story garage next door. Battalion Chief Joseph Stein requested an extra engine at 2:03 a.m. then called for Box 6318's full second alarm just one minute later. (MHFM, photograph by Henry Bryan.)

As civil unrest swept American cities in the days following Martin Luther King Jr.'s assassination, significant rioting, looting, and arson took place in Trenton on April 9–10, 1968. In the seven-hour period starting 6:25 p.m. on April 9, Trenton firefighters and mutual aid volunteer companies fought two dozen fires (including five major blazes) and answered nearly as many malicious false alarms. This image shows a rifle-carrying police officer on guard at the fire that gutted Convery's furniture store at 312 North Broad Street. Convery's furniture warehouse on Ringold Street was among the other properties torched. Fifteen firefighters were injured that night, some from being pelted with stones and bricks thrown by rioters. (MHFM, photograph by Warren Kruse.)

Reportedly caused by an oil burner explosion, this two-alarm blaze on July 12, 1968, badly damaged the City Tire Company at 909 South Broad Street. Firefighters were first alerted at 10:28 a.m., and Box 8373's second alarm was transmitted at 10:46 a.m. (MHFM, photograph by Warren Kruse.)

In the years after its 1961 opening to the public, the Trenton Fire Museum received donations of firefighting memorabilia from many supporters, including retired firefighters and their families. By 1969, the collection had outgrown its original space, so Chief Raymond Schwab offered up a new home for the museum—the much larger, no longer used gymnasium on Fire Headquarters' third floor. Firefighters are seen here in September 1969 preparing the new museum space for its public unveiling on October 5 during Fire Prevention Week. (MHFM, photograph by Henry Bryan.)

This two-alarm fire gutted the top two floors of Major's Furniture Mart at 144 North Broad Street on December 14, 1971. It was the first multiple-alarm fire since Chief Raymond Schwab died on October 4 (having suffered a heart attack during the three-alarm blaze on August 23 in the old Stern's department store at 124 North Broad Street). The first alarm for Major's was transmitted at 3:51 p.m. As heavy smoke billowed from the roof, Battalion Chief Francis Szmutko ordered Box 3113's second alarm at 3:54 p.m. Three days after the fire was out, like what happened in 1948 at Binder's (page 92), the interior of Major's collapsed without warning during the night. (MHFM.)

Breaking bones and suffering other injuries when the roof on which they were working collapsed, Capt. Thomas Walukiewicz and Stanley Chojnowski were among six firefighters hurt battling this general-alarm fire on February 19, 1972, that destroyed the warehouse at 119 Temple Street shared by Mercer Bag and Burlap and Convery's furniture companies. Both businesses had experienced prior devastating fires (pages 104, 112). Also damaged were the Pattern Machine and Foundry Company and five nearby homes. The blaze was reported at 9:43 a.m. Box 8319's fourth alarm was ordered at 10:54 a.m. Lawrence Township's Slackwood Fire Company, with its snorkel aerial apparatus, and other volunteer units were also called to the scene. (MHFM, photograph by John Pietras.)

Generating a massive smoke column, this two-alarm fire destroyed the abandoned Luzerne Rubber Company mill at 115 Muirhead Avenue on April 24, 1972. The first alarm was transmitted at 12:38 p.m., followed by Box 5314's second alarm at 12:41 p.m. Low water pressure hampered firefighting efforts, but flames were controlled before they could spread to the neighboring Star Porcelain Company. Two firefighters, Deputy Chief Vincent Pompei and Pasquale Migliaccio, suffered minor injuries. Over three decades earlier, on March 23, 1938, a three-alarm fire gutted a Luzerne Rubber Company warehouse on the same site. (MHFM, photograph by Warren Kruse.)

This two-alarm blaze, reported at 1:39 p.m. on February 1, 1973, and upgraded nine minutes later (Box 8321), damaged several classrooms at Trenton State Prison. Over the years, many fires have burned within the walls of the 1798 penitentiary house and its replacement, the fortress penitentiary built on the same site and opened to prisoners in 1836. Parts of the original prison burned in 1830 and 1835. More devastating fires struck the new prison on July 18, 1868, and September 8, 1880. Six inmates, having started the fire to aid their getaway, escaped during the 1868 blaze. Another two-alarm fire on November 23, 1936, gutted the prison bakery. On August 4, 1981, firefighters doused several small fires started during a prison riot. (MHFM, photograph by Martin D'Arcy.)

Chief Daniel P. George presents an award in the department's 1974 fire prevention essay contest. The annual contest in the 1970s involved elementary school writers explaining why people should not turn in false fire alarms. A World War II veteran, George joined the department in 1950. With his promotion in 1957 at age 30, he became the youngest-ever captain. He continued to set departmental "youngest" records with promotions to battalion chief (1964) and deputy chief (1966). Having served as acting chief since the death of Chief Raymond Schwab, George was sworn in as the department's 12th chief on December 24, 1971. He retired on April 30, 1992, as Trenton's longest-serving chief and died, aged 90, in 2017. (MHFM.)

Shown above at the 1975 general-alarm blaze that destroyed the Trenton Civic Center behind city hall is Ladder 4, a 1973 Ward LaFrance with an 85-foot Hi-Ranger snorkel that was reassigned in 1976 to Ladder 1. The fire's aftermath is below. Built between 1902 and 1905, the former National Guard armory was 270 feet long and 75 feet high. Dispatched at 11:56 p.m. on July 15, first-due firefighters ordered Box 6115's second alarm at 12:01 a.m. after encountering heavy smoke inside. The smoke and the building's massive size delayed finding the actual fire raging in a basement carpentry shop. The third alarm was called at 12:20 a.m., followed by the fourth at 12:36 a.m. Crews were ordered out when interior conditions deteriorated. Suburban volunteer companies were called to help with defensive operations. As flames broke through the roof and walls fell, billowing smoke carried embers for miles. Several firefighters suffered smoke inhalation. Irreplaceable city records stored inside burned with the building. The fire, widespread flooding that also happened that July, and the weeklong water crisis that resulted after valve failures flooded the city's filtration plant on August 31 made 1975's summer unforgettable for firefighters. (Above, MHFM, photograph by Warren Kruse; below, MHFM, photograph by Martin D'Arcy.)

The Plibrico Company, maker of heat-resistant bricks and other materials, was gutted by this three-alarm blaze on July 1, 1977. It was 4:46 p.m. when the first firefighters were sent to the factory at 1300 New York Avenue. Arriving to find heavy fire conditions, they requested Box 4515's second alarm at 4:52 p.m. The third alarm followed 20 minutes later. Firefighters shown in this photograph by Warren Kruse are wearing the Philadelphia-style helmets phased into the department in 1974–1975. These helmets did not last long and were replaced starting around 1987 by more traditional New York–style helmets like those previously used. (MHFM.)

As the second-largest fire in city history, topped only by Roebling's 1915 Buckthorn conflagration (page 63), this general-alarm inferno reported at 8:09 p.m. on March 1, 1979, raged out of control for over four hours and smoldered for days. Koenig and Sons' 700-foot-long plastics warehouse and several neighboring buildings occupied by the Kayline Processing plastics company were destroyed. Ironically, this fire off Lalor Street burned just a stone's throw from the site of the 1915 blaze. At least 15 volunteer companies helped fight the blaze. Hundreds were evacuated from their nearby homes due to toxic fumes generated by the burning plastics. (MHFM, photograph by Tom Herde.)

Accompanied by Chief Daniel George (left) and Deputy Chief Dennis Keenan (center), retired fire dispatcher Edwin Fisher sent in one final, ceremonial alarm on February 18, 1988, before Trenton's last street-corner fire alarm station, Box 2112 at Broad and State Streets, was taken out of service. Increasing false alarms (90 percent of 1986's 1,544 false alarms were turned in from street boxes), high maintenance costs, and improvements to the 911 telephone system prompted the city to begin removing all Gamewell alarm boxes in September 1987. (MHFM, photograph by Steve Mervish.)

Shown on October 3, 1988—the day Rescue 1 went in service as a company—are, from left to right, public safety director Richard Lucherini, Mayor Arthur J. Holland, Capt. Fred Marshall, Deputy Chief Dennis Keenan, and Chief Daniel George. Creation of this unit of men trained in technical rescue, hazardous materials, underwater recovery, and other specialties ended the decades-old practice of department rescue equipment, when needed, being cross-manned by different companies. Rescue 1 was assigned with Engine 1 and Ladder 1 to the 460 Calhoun Street firehouse, which opened on March 1, 1977, as a replacement for the old West Hanover Street firehouse, which closed on February 11, 1974, after it was declared structurally unsafe. (MHFM, photograph by Stacey Morgan.)

Five

1991–2020

This period is notable for two major, controversial changes made under Mayor Douglas H. Palmer.

The first of these, a restructuring of public safety services that was approved by voters in a special referendum on June 22, 1999, resulted in the uniformed posts of fire chief and police chief being abolished and replaced by new politically appointed civilian fire and police directors. Trenton's last fire chief, Robert C. Colletti, retired on July 12, 1999, the day Palmer's reforms took effect. Since then, the following have served as fire director, either full-time or in an acting role: Dennis M. Keenan, Richard J. Laird Jr., Henry E. Gliottone Jr., Leonard Carmichael Jr., Qareeb A. Bashir, Steven Coltre, and Derrick J.V. Sawyer. Each of these men, prior to his appointment as director, served as a uniformed Trenton firefighter and fire officer with the exception of Sawyer, a former Philadelphia firefighter and fire commissioner who came to Trenton in 2018.

The Palmer administration's other big shakeup was disbanding two engine companies and one ladder company, in part to reduce department operating and overtime expenses. Engine 2, Engine 5, and Ladder 3 all went permanently out of service on June 21, 2002.

The second firehouse expansion in recent years also took place during this period. Following the early 1980s building of an addition to Engine 3's South Broad Street quarters to accommodate Ladder 2 (with Ladder 2 relocating there on February 9, 1984, and its old South Clinton Avenue station later becoming home to Signal 22), construction of a large addition to Fire Headquarters on Perry Street started in October 1998. Engine 10 and Ladder 4 moved in on December 20, 2000, although work (including renovations to the old 1927 headquarters) continued for two more years.

On October 6, 2004, the Meredith Havens Fire Museum reopened to the public in a venue larger than ever—the refurbished first-floor apparatus bays of the old headquarters building.

A completely brand-new firehouse, the first since the Calhoun Street firehouse opened in 1977, was built at West State Street and Lee Avenue and formally dedicated on May 9, 2003, with Engine 9 relocating there from Brunswick Avenue a few days later. This change also resulted in Engine 6 moving from South Olden Avenue into Ladder 3's old quarters on North Clinton Avenue.

In 2007, the department was given a new name—Trenton Department of Fire and Emergency Services—and all members began receiving first responder training to respond to medical emergencies in support of Trenton Emergency Medical Service (TEMS) ambulance staff.

In January 2020, the department began to be featured on the weekly A&E reality television show *Live Rescue*. Among the department's responses documented on the show was a general-alarm blaze on Brunswick Avenue on Super Bowl Sunday, February 2, 2020.

Trenton's front-line apparatus and various support equipment are shown here prior to the April 4, 1992, parade that celebrated the 100th anniversary of the paid department. The procession, led by the department honor guard (organized 1987), recreated the volunteer firefighters' farewell march of a century earlier. Many of the rigs are painted the lime yellow the department used in the 1970s and early 1980s. The newer engines are KMEs; the older ones are Maxims, FTIs, a Hahn, and a Sutphen. The tillered ladders are American LaFrances and a Pirsch. The tower ladder is a Thibault Skypod. Rescue 1's rig is a 1987 GMC/Emergency One. (MHFM.)

Shown in May 1992 are, from left to right, Chief Dennis M. Keenan, Deputy Chief Robert C. Colletti, and Capt. Sidney Skinner. Joining the department in 1962, Keenan was promoted to captain in 1969, battalion chief in 1974, and deputy chief in 1982. Following Chief Daniel George's retirement, Keenan was sworn in as the department's 13th chief on May 1, 1992. After Keenan retired on June 1, 1998, to become Trenton's new public safety director, Colletti became acting chief. Colletti was officially promoted on October 26, 1998, as the 14th and final chief. On the job since 1964, Colletti retired on July 12, 1999, after the chief's position was abolished. (MHFM.)

Multiple fires set by an arsonist developed into this September 4, 1996, inferno that destroyed four abandoned buildings in the former Kramer Trenton Company complex on North Olden Avenue where heating and air-conditioning equipment were once made. Dispatched at 8:38 p.m., Battalion Chief Thomas Murl and first-due firefighters found a 450-foot-long structure well involved in fire, with flames rapidly spreading. The fourth alarm, calling in the last of Trenton's on-duty firefighters, was struck at 9:25 p.m. Four more alarms of apparatus and manpower from volunteer companies from Mercer County and neighboring Bucks County, Pennsylvania, were then requested to help supply water and protect nearby homes and businesses. (Photograph by Michael Ratcliffe.)

One person was killed in this fire on April 12, 1997, at 144 Hoffman Avenue. Alerted at 1:05 a.m., firefighters found heavy fire venting from the upper floors of the boarded-up house. Crews from Engines 8, 1, and 5, Ladders 1 and 4, and Rescue 1 directed by Battalion Chief Chester Haymond forced entry and mounted an aggressive interior attack. Once inside, they located the body of a 49-year-old homeless man who had been squatting in the structure. (Photograph by Michael Ratcliffe.)

Three alarms were transmitted for this blaze at 29–33 Oxford Street on November 4, 1998. Engine 5 firefighters, investigating reports of smoke in the area, discovered the fire in the vacant warehouse at 4:13 p.m. The rest of the first-alarm assignment—Engines 10 and 1, Ladders 4 and 1, and Rescue 1—were promptly dispatched. Engines 6 and 9 were sent at 4:30 p.m. after Battalion Chief Chester Haymond called the second alarm. Engines 2 and 3 and Ladder 3 responded at 4:45 p.m. after the third alarm was struck. The fire was declared under control at 5:57 p.m. (Photograph by Michael Ratcliffe.)

Silhouetted by flames venting behind them, members of Ladder 4 quickly evacuate the roof of an exposure building. The fire in a vacant, boarded-up structure in the 300 block of Centre Street was reported at 10:22 p.m. on May 22, 2000. Four engine companies, two ladder companies, and Rescue 1 fought the blaze and had it under control by 11:10 p.m. (Photograph by Michael Ratcliffe.)

Less than three weeks after Engines 2 and 5 and Ladder 3 were disbanded, the apparatus previously assigned to those companies briefly returned to action, manned by recalled off-duty personnel, to help battle flames and cover the city during this general-alarm fire at Trenton Psychiatric Hospital on July 9, 2002. The blaze, reported at 7:54 a.m., consumed the roof of the Haines Building on the hospital's Sullivan Way campus. Thirteen volunteer companies from Ewing, Lawrence, Hamilton, and other towns helped with water supply and firefighting operations. Shown here with its Readi-Tower squirt boom elevated and flowing water is Engine 1. (Photograph by Michael Ratcliffe.)

A 21-year-old woman and her young sons, aged 2 and 3, died in this two-alarm fire that charred five rowhouses at 29–37 Garfield Avenue on July 26, 2004. Alerted at 5:00 a.m., firefighters arrived to find the victims' home already engulfed in flames. Investigators later determined flames had burned undetected for a long time, trapping the three because their home had no working smoke detectors to warn them. After this tragedy, firefighters went door-to-door to stress the importance of smoke detectors and install free detectors in many homes. (Photograph by Michael Ratcliffe.)

Over the years, department members have made many dramatic rescues that did not involve any actual firefighting, like the April 5, 2002, saving of a suicidal man threatening to jump 14 stories from the sign atop the Broad Street Bank building. Shown here is another example of firefighters' ever-expanding lifesaver role—a trench rescue on South Hermitage Avenue on May 4, 2007. Aided by other emergency personnel, the department's technical rescue task force (Rescue 1, Ladder 4, and Engine 10) worked for 90 minutes to successfully free a plumbing contractor who became trapped up to his chest in dirt and had his right leg crushed under a heavy slab of concrete when the walls of a six-foot-deep hole caved in around him. (Photograph by Michael Ratcliffe.)

This January 30, 2008, photograph shows the expanded Trenton Fire Headquarters after new LEDs were installed to illuminate its distinctive sign and accompanying 12-by-25-foot firefighters' helmet icon. After the headquarters addition opened in 2000 (the helmet being added in 2002), the sign's original neon lighting system was plagued by electrical problems that often darkened random letters and at other times left the entire sign unlighted. (Photograph by Michael Ratcliffe.)

The *Iron Fireman*—a recognized symbol of the Trenton Fire Department for well over a century that prominently appears on the department's uniform patch—stands in front of Trenton City Hall atop a base adorned with plaques bearing the names of firefighters from Trenton and other Mercer County towns who have died in the line of duty. The statue, made by J.W. Fiske of New York, was originally dedicated on July 28, 1892, in memory of the city's disbanded volunteer department. Erected outside the old city hall at State and Broad Streets, it stood on a base featuring a pedestrian drinking fountain and watering trough for horses. In 1911, the statue and its fountain base were relocated to the Stockton Street side of the new city hall built on East State Street. The *Iron Fireman* was removed in 1960 after cracks started appearing in its old fountain base. Following refurbishment financed by the city firefighters' union, on April 8, 1962, the statue was rededicated on a new granite base in its current location in front of city hall. Badly damaged by a vandal in April 2013, the statue was repaired and again rededicated on September 9, 2018. (Photograph by Michael Ratcliffe.)

ROLL OF HONOR

Thirty-four Trenton firefighters, paid and volunteer, are known to have lost their lives in the line of duty. This list is as complete as possible given the available information. But there may be others who made the ultimate sacrifice missing from this roll. Possible reasons for any such omission are many—poor documentation, lost records, outdated criteria that restricted line-of-duty recognition only to those deaths occurring in a very limited set of circumstances, and the stoic but misguided view of past generations of firefighters that disfiguring injuries and death were a part of the job that had to be accepted. This list was compiled from department records and newspaper accounts. The abbreviation Ff. is used here for the rank of firefighter.

1864—**Ff. Robert S. Anthony** (Good Will), 26, was crushed to death on May 20 when, while returning from a fire, he tripped and fell beneath the wheels of the steam fire engine he was helping to pull.

1883—**Foreman Ephraim J. Dollas** (Union), 32, died on May 13 of an inflammatory illness contracted as he led his company during a fire at the Mansion House, 15–17 East State Street, on May 8, 1883.

1883—**Ff. Charles J. Bates** (Eagle), 35, died on October 24 from the lasting effects of a severe cold he caught while battling a fierce blaze in a century-old gristmill during a February 4, 1882, snowstorm.

1884—**Past Asst. Chief Engineer George W. Vanhorn** (Ossenberg), 30, died on October 28 from tuberculosis brought on by a cold he caught as he directed firefighting efforts during the September 30, 1883, blaze that destroyed St. John's Catholic Church on South Broad Street, near Centre Street.

1885—**Ff. Thomas McGowan** (Delaware), 26, died on March 9 of injuries he received three days prior when he was thrown from his horse-drawn apparatus as it crashed going to a fire in a shoe store.

1885—**Ff. Joseph T. Kinney** (Eagle), 26, died on April 8 of a lasting cold he contracted while fighting the December 7, 1884, fire that destroyed the Home Rubber Company's factory in Chambersburg.

1885—**Foreman Maurice "Morris" H. Ely** (Ossenberg), 32, died on October 18 from tuberculosis that developed after he helped fight, in frigid weather, the March 21, 1885, fire at the state house.

1888—**Ff. Harry A. Stradling** (Hand-In-Hand), 29, was fatally injured on April 29 when he fell out of an opening in the Hand-In-Hand firehouse hay loft and landed on the cobblestones 12 feet below.

1896—**Ff. Charles B. Wood** (Engine 6), 49, was fatally crushed beneath a wall that collapsed during the general-alarm fire at Trenton Fire Clay and Porcelain Company off Third Street on August 6.

1899—**Capt. James Nugent** (Engine 1), 61, died on May 24 of a heart attack he suffered while on duty in his firehouse just hours after he led his men into a smoky first-due blaze in a tobacco store.

1899—**Ff. Charles H. Tindall** (Truck 2), 30, died on July 19 of tuberculosis aggravated by chest injuries he suffered on March 11, 1896, when Truck 2, while going to a fire, crashed into a trolley car.

1899—**Ff. John P. Henry** (Engine 3), 39, died on August 14 of complications of a surgery needed after one of his legs was badly crushed between Engine 3's steamer and hose wagon on June 28, 1899.

1901—**Ff. John McGowan** (Truck 1), 47, died on November 21, seventeen days after his skull was crushed by falling bricks at a general-alarm fire (and 16 years after his brother Thomas's line-of-duty death).

1906—**Ff. Charles A. Howell** (Chemical 1), 32, and **Ff. Frank C. Riley** (Engine 4), 30, were crushed to death on May 12 when, during a two-alarm fire on New York Avenue, the New Jersey Pulp Plaster Company building collapsed on top of them. Howell was posthumously promoted to lieutenant.

1907—**Ff. Edward H. Hizer** (Truck 1), 55, was killed on March 29 when, due to a pothole in the road, he was tossed from the driver's seat and run over by his own truck while responding to a barn fire.

1908—**Ff. Frank Schollenberger** (Engine 7), 49, died on November 28 of a heart attack that—brought on by injuries suffered at a fire a week earlier—struck him as he went to light his firehouse's boiler.

1915—**Lt. Asa Lanning** (Engine 6), 43, died on April 27 from the lingering effects of injuries suffered on May 12, 1912, when his horse-drawn apparatus crashed into a shed while heading to a fire.

1915—**Ff. Frederick G. Slover** (Engine 5), 29, also died on April 27 when a wall collapsed on him as he manned a hoseline during a three-alarm fire at the American Rubber plant on Perrine Avenue.

1916—**Capt. Alexander Grugan** (Truck 3), 55, died on December 13, just one day after he suffered a stroke while on duty and was badly injured in a head-first fall down his firehouse's stairs.

1918—**Ff. John E. Owens** (Engine 8), 59, died of a heart attack on January 7, shortly after he collapsed from smoke inhalation while fighting a fire at New Jersey School and Church Furniture Company.

1936—**Ff. Fritz F. Schmiedel** (Engine 10), 44, died on May 1 of a heart attack suffered while on duty.

1936—**Deputy Chief Augustus W. Conway**, 49, died on November 16 of an on-duty heart attack.

1951—**Ff. George Weeks** (Engine 1), 47, died on February 26, less than 48 hours after he was thrown from the rear step of Engine 1, fracturing his skull, when the rig was struck by a passenger car.

1952—**Capt. John H. Sawyer** (Engine 6), 57, died of a heart attack while on duty in his firehouse.

1953—**Ff. John J. O'Connell** (Engine 1), 53, died of a massive heart attack on April 30 as he manned Engine 1's pump panel during a smoky blaze involving two houses at 96–98 Belvidere Street.

1956—**Capt. Richard C. Green** (Engine 2), 42, was killed on February 11 when, slipping off an aerial ladder, he fell 50 feet while fighting a fire on the roof of the Hotel Penn, 81 South Clinton Avenue.

1966—**Ff. Peter P. DelAversano Jr.** (Engine 10), 24, was killed on September 10 when Engine 2 and Engine 10, both responding to a two-alarm fire, collided at Chambers Street and Hamilton Avenue.

1971—**Chief Raymond A. Schwab**, 61, died on October 4 of cardiac failure brought on by the heart attack he suffered while directing firefighting operations at a three-alarm blaze on August 23, 1971.

1974—**Battalion Chief Thomas W. Mayer**, 48, died on January 2, the result of a heart attack suffered the previous evening while he supervised firefighters battling a blaze at 417 Monmouth Street.

1982—**Ff. William A. Womack Jr.** (Engine 7), 31, died on June 18 shortly after he collapsed, reportedly from bronchial pneumonia, during pump operator training at Canal Boulevard off Lalor Street.

1986—**Ff. Robert J. Mizopalko** (Ladder 4), 33, and **Ff. Joseph F. Woods Jr.** (Engine 10), 25, died on August 4 after heavy smoke caused them to become disoriented and lost and they ran out of air in their breathing apparatus during a three-alarm fire in Shenanigan's Saloon on South Warren Street.

2009—**Ff. Manuel "Manny" Rivera Sr.** (Engine 3), 42, died on March 31 of injuries suffered on February 9 as he (while on an overtime shift with Engine 7) rescued a man from a fire on Washington Street.

DISCOVER THOUSANDS OF LOCAL HISTORY BOOKS FEATURING MILLIONS OF VINTAGE IMAGES

Arcadia Publishing, the leading local history publisher in the United States, is committed to making history accessible and meaningful through publishing books that celebrate and preserve the heritage of America's people and places.

Find more books like this at
www.arcadiapublishing.com

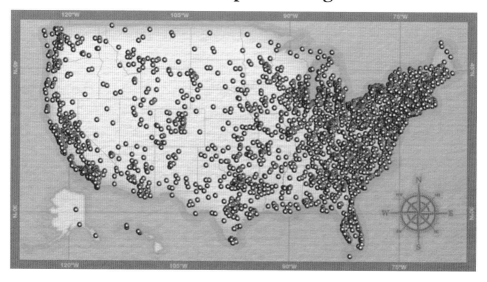

Search for your hometown history, your old stomping grounds, and even your favorite sports team.

Consistent with our mission to preserve history on a local level, this book was printed in South Carolina on American-made paper and manufactured entirely in the United States. Products carrying the accredited Forest Stewardship Council (FSC) label are printed on 100 percent FSC-certified paper.

MADE IN THE USA